$1-\cos\theta = 2\pi$

21.2.

MECHANICS
AND
PROPERTIES OF MATTER

BOOKS BY M. NELKON

Published by Heinemann
OPTICS, SOUND AND WAVES (SI)
SCHOLARSHIP PHYSICS (SI)
ADVANCED LEVEL PHYSICS (SI) (*with P. Parker*)
MECHANICS AND PROPERTIES OF MATTER (SI)
PRINCIPLES OF ATOMIC PHYSICS AND ELECTRONICS (SI)
ADVANCED LEVEL PRACTICAL PHYSICS (SI) (*with J. Ogborn*)
AN INTRODUCTION TO THE MATHEMATICS OF PHYSICS (SI)
(*with J. H. Avery*)
ELEMENTARY PHYSICS (SI) (*with A. F. Abbott*)
GRADED EXERCISES AND WORKED EXAMPLES IN PHYSICS (SI)
ELECTRONICS AND RADIO (SI) (*with H. I. Humphreys*)
NEW TEST PAPERS IN PHYSICS (SI)
REVISION NOTES IN PHYSICS (SI)
Book I. Mechanics, Electricity, Atomic Physics
Book II. Optics, Waves, Sound, Heat, Properties of Matter
SOLUTIONS TO ORDINARY LEVEL PHYSICS QUESTIONS (SI)
SOLUTIONS TO ADVANCED LEVEL PHYSICS QUESTIONS (SI)
REVISION BOOK IN ORDINARY LEVEL PHYSICS (SI)

Published by Arnold
ELECTRICITY (Advanced Level-SI)

Published by Blackie
HEAT (Advanced Level-SI)

Published by Chatto and Windus
PRINCIPLES OF PHYSICS (Ordinary Level-SI)
EXERCISES IN ORDINARY LEVEL PHYSICS (SI)
C.S.E. PHYSICS (SI)
REVISION BOOK IN BASIC PHYSICS (SI)
SI UNITS: AN INTRODUCTION FOR A-LEVEL

MECHANICS AND PROPERTIES OF MATTER

by

M. NELKON, M.Sc. (Lond.), F.Inst.P., A.K.C.
Formerly Head of the Science Department, William Ellis School, London

FOURTH EDITION WITH SI UNITS

HEINEMANN EDUCATIONAL BOOKS
LONDON

Heinemann Educational Books Ltd
LONDON   EDINBURGH   MELBOURNE   AUCKLAND   TORONTO
SINGAPORE   HONG KONG   KUALA LUMPUR   IBADAN
NAIROBI   JOHANNESBURG   NEW DELHI   LUSAKA

ISBN 0 435 68633 X
© M. Nelkon 1969

First published 1952
Second edition 1955
Reprinted 1955, 1956, 1957
Third edition 1958
Reprinted 1959, 1960, 1961, 1963, 1965
Reprinted with additions 1966
Fourth edition 1969 (Reset)
Reprinted 1970, 1972, 1974 with corrections

Published by
Heinemann Educational Books Ltd
48 Charles Street, London W1X 8AH

Filmset by Keyspools Ltd., Golborne, Lancashire
Printed in Great Britain by
Butler & Tanner Ltd., Frome, Somerset

NOTE TO 1974 REPRINT

IN this reprint, I have taken the opportunity to revise the text further in SI units and to correct some misprints. I am particularly indebted to Dr. Rai, formerly head of the Science Department of National Council of Education and Research, New Delhi, India, F. Anstis, Reed's School, Surrey, and Dr. P. E. Chandler, King's School, Gloucester, for their helpful comments.

PREFACE TO FOURTH EDITION

IN this edition, I have taken the opportunity to revise and re-order the whole text. Among the changes are: (i) introduction of SI units, (ii) further treatment of linear momentum, (iii) emphasis on centripetal force, (iv) addition of orbits and satellites in gravitation, (v) discussion of waves, energy changes in simple harmonic motion, and damped oscillations, (vi) emphasis of conservation of angular momentum, (vii) consideration of buoyancy in accelerated liquids, (viii) discussion of intermolecular forces and states of matter, (ix) further account of surface energy, (x) omission of osmosis and diffusion, which are discussed in physical chemistry texts. The exercises at the end of chapters have also been revised and recent examination questions added. Revision papers, containing miscellaneous questions, are now included at the end of the book.

In connection with the revision, I am particularly grateful for assistance by S. M. Freake, Queen's College, Cambridge, R. P. T. Hills, St. John's College, Cambridge, and M. V. Detheridge, Woodhouse Grammar School, London; and for constructive criticisms particularly to Mrs. A. Bradshaw, City of London Girls School; G. Ullyott, Charterhouse School; R. D. Harris, Ardingley College, Sussex; and A. C. Covell, Rickmansworth Grammar School.

I also acknowledge with thanks the permission to reprint questions set in past Advanced level examinations by the following Examining Boards. I am particularly indebted to the five Boards first named for permission to translate numerical values in past questions to SI units; the translation is the sole responsibility of the author.

London University Schools Examination Council	(L.)
Joint Matriculation Board	(N.)
Oxford and Cambridge Joint Board	(O. & C.)
Cambridge Local Examinations Syndicate	(C.)
Oxford Local Examinations	(O.)
Welsh Joint Education Committee	(W.)

PREFACE TO FIRST EDITION

THIS text-book is written for General Certificate Advanced level and Special level students, and covers, to that standard, Dynamics, Statics, Fluids, Surface Tension, Elasticity, Friction in Solids, and Viscosity. The book is based on the lessons and lectures given to sixth-form and intermediate students at schools and polytechnics, and in the treatment a knowledge of Physics to the General Certificate Ordinary level has been assumed.

I have begun Mechanics with dynamics, circular and simple harmonic motion, and moments of inertia, as these topics have more applications in Physics than statics; the theory of dimensions has been utilised where the mathematics is too difficult, as in the subject of viscosity for example; the 'excess pressure' formula has been extensively used in the treatment of surface tension; and worked examples have been added to the text to assist the student. The exercises at the end of chapters are graded, and there are appendices containing basic definitions and formulae to help revision. Accounts of the history of mechanics have been omitted as this is usually treated at length in text-books of Ordinary level, and the kinetic theory of gases is discussed in another volume on Heat. I have endeavoured to explain clearly the physical principles of the subjects, and it is hoped that the book will prove a useful introduction and stepping-stone to the many existing text-books of higher standard in the subject.

I acknowledge with thanks the generous assistance with the work given to me by P. Parker, M.Sc., late Senior Lecturer, City University, London, A. W. K. Ingram, M.A., Senior Science Master, Lawrence Sheriff School, Rugby, C. R. Ensor, M.A., Senior Physics Master, Downside School, Bath, and D. W. Stops, B.Sc., Ph.D., Reader, City University, London, who kindly read the proofs and made suggestions.

CONTENTS

Chapter *Page*

1. DYNAMICS ... 1
   Motion in straight line. Vectors, scalars. Velocity-time graph. Components. Motion under gravity. Vector addition and subtraction. Newton's Laws. Force. Linear momentum. Work. Kinetic, potential energy. Conservative, non-conservative forces. Einstein's mass-energy relation. Dimensions.

2. CIRCULAR MOTION. SIMPLE HARMONIC MOTION. GRAVITATION ... 40
   Acceleration in circle, centripetal forces. Variation of g. S.H.M. Damping. Simple pendulum, spiral spring, liquid in U-tube. Potential and kinetic energy exchanges. Waves-progressive, stationary, speed, Gravitation. Kepler's laws. G. Orbits, weightlessness. Values of g. Velocity of escape.

3. ROTATION OF RIGID BODIES ... 86
   Kinetic energy. Moment of inertia of rod, ring, disc, cylinder, sphere. Theorems of parallel and perpendicular axes. Couple and angular acceleration. Angular momentum and conservation. Energy of rolling object. Period of oscillation. Moment of inertia determination. Compound pendulum.

4. STATIC BODIES. FLUIDS ... 110
   Static Bodies. Parallelogram and triangle of forces. Moments. Parallel forces. Equilibrium of three forces. Moment of couple. Centre of gravity and of mass. Types of equilibrium. Common balance and sensitivity. **Fluids.** Pressure in liquid. Atmospheric pressure, variation of height. Brownian motion. Density, relative density. Archimedes' Principle. Flotation. Hydrometer. Accelerated liquids. Bernoulli's theorem and applications.

5. SURFACE TENSION ... 145
   Intermolecular forces. States of matter. Surface tension phenomena. Definition, units. Capillarity. Angle of contact. Measurement by capillary tube and balance. Excess pressure in bubble and curved liquid surface. Capillary rise and fall. Temperature variation of surface tension. Cylindrical surfaces, falling drop. Surface energy. Solid-liquid, vapour-liquid interfaces

6. ELASTICITY 180
   Elastic limit. Hooke's law. Yield point. Ductile, brittle substances. Tensile stress and strain. Young's modulus and determination. Force in heated or cooled bar. Energy in stretched wire. Bulk modulus in liquids and gases. Modulus of rigidity. Torsion wire. Poisson's ratio.

7. SOLID FRICTION. VISCOSITY 200
   Static and kinetic friction. Coefficient of friction. Theory of friction. Viscosity of liquids. Coefficient of viscosity. Liquid flow. Poiseuille's formula. Laminar, turbulent flow. Viscometer. Stokes' law. Terminal velocity. Measurement by falling sphere. Rotating cylinder. Molecular theory of viscosity.

   REVISION PAPERS 223

   ANSWERS 229

   INDEX 233

MECHANICS AND PROPERTIES OF MATTER ERRATUM

Please note that the answers to these exercises are given incorrectly in the 1974 reprint of this book.

The following are the correct answers:

EXERCISES 1 (p. 35)
- **25.** 1667 N; 83,330 J.
- **32.** 8000 N, 22 kW.

EXERCISES 2 (p. 80)
- **10.** A.
- **11.** (i) 2 rad s^{-1}, (ii) 96 N.
- **12.** (i) 118 N, (ii) 32°.
- **13.** 42°, 13,450 N.
- **14.** 224 N, 64 N.

EXERCISES 5 (p. 175)
- **16.** (ii) 9×10^{-3} J.

EXERCISES 6 (p. 196)
- **15.** 70 N.
- **18.** 117 g.
- **24.** 144°C, 33,000 N.
- **29.** 280 N.
- **30.** 12×10^7 N m^{-2}, 3 kg.

Chapter 1

DYNAMICS

Motion in a Straight Line. Velocity

IF a car travels steadily in a constant direction and covers a distance s in a time t, then its *velocity* in that direction $= s/t$. If the car does not travel steadily, then s/t is its average velocity, and

$$\text{distance } s = \text{average velocity} \times t.$$

We are here concerned with motion in a constant direction. The term 'displacement' is given to the distance moved in a constant direction, for example, from L to C in Fig. 1.1 (i). Velocity may therefore be defined as the *rate of change of displacement*.

Velocity can be expressed in *centimetres per second* (cm/s or cm s^{-1}) or *metres per second* (m/s or m s^{-1}) or *kilometres per hour* (km/h or km h^{-1}). By calculation, 36 km h^{-1} = 10 m s^{-1}. It should be noted that complete information is provided for a velocity by stating its direction in addition to its magnitude, as explained shortly.

If an object moving in a straight line travels equal distances in equal times, no matter how small these distances may be, the object is said to be moving with *uniform* velocity. The velocity of a falling stone increases continuously, and so is a *non-uniform* velocity.

If, at any point of a journey, Δs is the small change in displacement in a small time Δt, the velocity v is given by $v = \Delta s/\Delta t$. In the limit, using calculus notation,

$$v = \frac{ds}{dt}.$$

Vectors

Displacement and *velocity* are examples of a class of quantities called *vectors* which have both magnitude and direction. They may therefore be represented to scale by a line drawn in a particular direction. Thus

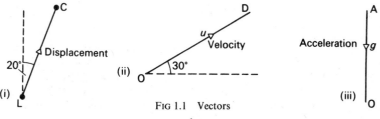

FIG 1.1 Vectors

1

Cambridge is 80 km from London in a direction 20° E. of N. We can therefore represent the displacement between the cities in magnitude and direction by a straight line LC 4 cm long 20° E. of N., where 1 cm represents 20 km, Fig. 1.1 (i). Similarly, we can represent the velocity u of a ball initially thrown at an angle of 30° to the horizontal by a straight line OD drawn to scale in the direction of the velocity u, the arrow on the line showing the direction, Fig. 1.1 (ii). The acceleration due to gravity, g, is always represented by a straight line AO to scale drawn vertically downwards, since this is the direction of the acceleration, Fig. 1.1 (iii). We shall see later that 'force' and 'momentum' are other examples of vectors.

Speed and Velocity

A car moving along a winding road or a circular track at 80 km h^{-1} is said to have a *speed* of 80 km h^{-1}. 'Speed' is a quantity which has no direction but only magnitude, like 'mass' or 'density' or 'temperature'. These quantities are called *scalars*.

The distinction between speed and velocity can be made clear by reference to a car moving round a circular track at 80 km h^{-1} say. Fig. 1.2. At every point on the track the *speed* is the same—it is 80 km h^{-1}.

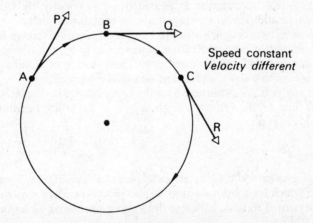

FIG. 1.2. Velocity and speed

At every point, however, the *velocity* is different. At A, B or C, for example, the velocity is in the direction of the particular tangent, AP, BQ or CR, so that even though the magnitudes are the same, the three velocities are all different because they point in different directions. Generally, vector quantities can be represented by a line drawn in the direction of the vector and whose length represents its magnitude.

Distance-Time Curve

When the displacement, or distance, s of a moving car from some fixed point is plotted against the time t, a *distance-time* ($s-t$) *curve* of the motion is obtained. The velocity of the car at any instant is given by the change in distance per second at that instant. At E, for example, if the change in distance s is Δs and this change is made in a time Δt,

$$\text{velocity at E} = \frac{\Delta s}{\Delta t}.$$

In the limit, then, when Δt approaches zero, the velocity at E becomes equal to the *gradient of the tangent to the curve* at E. Using calculus notation, $\Delta s/\Delta t$ then becomes equal to ds/dt (p. 1).

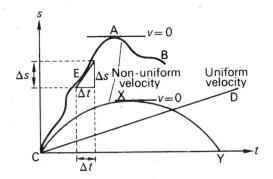

FIG. 1.3   Displacement (s)–time (t) curves

If the distance-time curve is a straight line CD, the gradient is constant at all points; it therefore follows that the car is moving with a *uniform* velocity, Fig. 1.3. If the distance-time curve is a curve CAB, the gradient varies at different points. The car then moves with non-uniform velocity. We may deduce that the velocity is zero at the instant corresponding to A, since the gradient at A to the curve CAB is zero.

When a ball is thrown upwards, the height s reached at any instant t is given by $s = ut - \tfrac{1}{2}gt^2$, where u is the initial velocity and g is the constant equal to the acceleration due to gravity (p. 8). The graph of s against t is represented by the parabolic curve CXY in Fig. 1.3; the gradient at X is zero, illustrating that the velocity of the ball at its maximum height is zero.

Velocity-Time Curves

When the velocity of a moving train is plotted against the time, a 'velocity-time (v-t) curve' is obtained. Useful information can be deduced from this curve, as we shall see shortly. If the velocity is uniform, the velocity-time graph is a straight line parallel to the time-axis, as shown by line (1) in Fig. 1.4. If the train accelerates uniformly from rest, the velocity-time graph is a straight line, line (2), inclined to the time-axis. If the acceleration is not uniform, the velocity-time graph is curved.

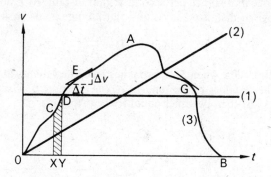

FIG. 1.4 Velocity (v)–time (t) curves

In Fig. 1.4, the velocity-time graph OAB represents the velocity of a train starting from rest which reaches a maximum velocity at A, and then comes to rest at the time corresponding to B; the acceleration and retardation are both not uniform in this case.

Acceleration is the 'rate of change of velocity', i.e. the change of velocity per second. *The acceleration of the train at any instant is given by the gradient to the velocity-time graph* at that instant, as at E. At the peak point A of the curve OAB the gradient is zero, i.e., the acceleration is then zero. At any point, such as G, between A, B the gradient to the curve is negative, i.e., the train undergoes retardation.

The gradient to the curve at any point such as E is given by:

$$\frac{\text{velocity change}}{\text{time}} = \frac{\Delta v}{\Delta t}$$

where Δv represents a small change in v in a small time Δt. In the limit, the ratio $\Delta v/\Delta t$ becomes dv/dt, using calculus notation.

Area Between Velocity-Time Graph and Time-Axis

Consider again the velocity-time graph OAB, and suppose the velocity increases in a very small time-interval XY from a value represented by XC to a value represented by YD, Fig. 1.4. Since the small distance travelled = average velocity × time XY, the distance travelled is represented by the *area* between the curve CD and the time-axis, shown shaded in Fig. 1.4. By considering every small time-interval between OB in the same way, it follows that *the total distance travelled* by the train in the time OB is given by the area between the velocity-time graph and the time-axis. This result applies to any velocity-time graph, whatever its shape.

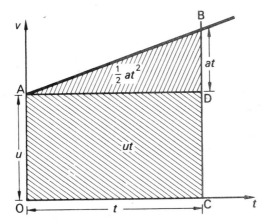

FIG. 1.5 Uniform acceleration

Fig. 1.5 illustrates the velocity-time graph AB of an object moving with uniform acceleration a from an initial velocity u. From above, the distance s travelled in a time t or OC is equivalent to the area OABC. The area OADC $= u \cdot t$. The area of the triangle ABD $= \frac{1}{2}$AD $\cdot$ BD $= \frac{1}{2}t \cdot$ BD. Now BD $=$ the increase in velocity in a time t $= at$. Hence area of triangle ABD $= \frac{1}{2}t \cdot at = \frac{1}{2}at^2$

$$\therefore \text{total area OABC} = s = ut + \tfrac{1}{2}at^2.$$

This result is also deduced on p. 6.

Acceleration

The *acceleration* of a moving object at an instant is the *rate of change of its velocity* at that instant. In the case of a train accelerating steadily from 36 km h^{-1} (10 m s^{-1}) to 54 km h^{-1} (15 m s^{-1}) in 10 second, the uniform acceleration

$$= (54-36) \text{ km h}^{-1} \div 10 \text{ seconds} = 1 \cdot 8 \text{ km h}^{-1} \text{ per second,}$$

or

$$(15-10) \text{ m s}^{-1} \div 10 \text{ seconds} = 0 \cdot 5 \text{ m s}^{-1} \text{ per second.}$$

Since the time element (second) is repeated twice in the latter case, the acceleration is usually given as $0 \cdot 5$ m s^{-2}. Another unit of acceleration is 'cm s^{-2}'. In terms of the calculus, the acceleration a of a moving object is given by

$$a = \frac{dv}{dt}$$

where dv/dt is the velocity change per second.

Distance Travelled with Uniform Acceleration. Equations of Motion

If the velocity changes by equal amounts in equal times, no matter how small the time-intervals may be, the acceleration is said to be *uniform*. Suppose that the velocity of an object moving in a straight line with uniform acceleration a increases from a value u to a value v in a time t. Then, from the definition of acceleration,

$$a = \frac{v-u}{t},$$

from which $\qquad \mathbf{v = u + at}$ . . . . (1)

Suppose an object with a velocity u accelerates with a uniform acceleration a for a time t and attains a velocity v. The distance s travelled by the object in the time t is given by

$$s = \text{average velocity} \times t$$
$$= \tfrac{1}{2}(u+v) \times t$$

But $\qquad v = u + at$

$$\therefore s = \tfrac{1}{2}(u + u + at)t$$

$$\therefore \mathbf{s = ut + \tfrac{1}{2}at^2} \quad . \quad . \quad . \quad . \quad (2)$$

If we eliminate t by substituting $t = (v-u)/a$ from (1) in (2), we obtain, on simplifying,

$$\mathbf{v^2 = u^2 + 2as} \quad . \quad . \quad . \quad (3)$$

Equations (1), (2), (3) are the equations of motion of an object moving in a straight line with uniform acceleration. When an object undergoes a uniform *retardation*, for example when brakes are applied to a car, a has a *negative* value.

EXAMPLES

1. A car moving with a velocity of 54 km h^{-1} accelerates uniformly at the rate of 2 m s^{-2}. Calculate the distance travelled from the place where acceleration began to that where the velocity reaches 72 km h^{-1}, and the time taken to cover this distance.

(i) 54 km h^{-1} = 15 m s^{-1}, 72 km h^{-1} = 20 m s^{-1}, acceleration a = 2 m s^{-2}.

Using
$$v^2 = u^2 + 2as,$$
$$\therefore 20^2 = 15^2 + 2 \times 2 \times s$$
$$\therefore s = \frac{20^2 - 15^2}{2 \times 2} = 43\tfrac{3}{4} \text{ m.}$$

(ii) Using
$$v = u + at$$
$$\therefore 20 = 15 + 2t$$
$$\therefore t = \frac{20 - 15}{2} = 2\cdot 5 \text{ s.}$$

2. A train travelling at 72 km h^{-1} undergoes a uniform retardation of 2 m s^{-2} when brakes are applied. Find the time taken to come to rest and the distance travelled from the place where the brakes were applied.

(i) 72 km h^{-1} = 20 m s^{-1}, and $a = -2$ m s^{-2}, $v = 0$.

Using
$$v = u + at$$
$$\therefore 0 = 20 - 2t$$
$$\therefore t = 10 \text{ s}$$

(ii) The distance, $s, = ut + \tfrac{1}{2}at^2$.
$$= 20 \times 10 - \tfrac{1}{2} \times 2 \times 10^2 = 100 \text{ m.}$$

Motion Under Gravity.

When an object falls to the ground under the action of gravity, experiment shows that the object has a constant or uniform acceleration of about 980 cm s^{-2}, while it is falling (see p. 48). In SI units this is 9·8 m s^{-2} or 10 m s^{-2} approximately. The numerical value of this acceleration is usually denoted by the symbol g. Suppose that an object is dropped from a height of 20 m above the ground. Then the initial velocity $u = 0$, and the acceleration $a = g = 10$ m s^{-2} (approx). Substituting in $s = ut + \tfrac{1}{2}at^2$, the distance fallen s in metres is calculated from

$$s = \tfrac{1}{2}gt^2 = 5t^2.$$

When the object reaches the ground, $s = 20$ m.
$$\therefore 20 = 5t^2, \text{ or } t = 2 \text{ s}$$

Thus the object takes 2 seconds to reach the ground.

If a cricket-ball is thrown vertically upwards, it slows down owing to the attraction of the earth. The ball is thus retarded. The magnitude of the retardation is 9·8 m s^{-2}, or g. Mathematically, a retardation can be regarded as a negative acceleration in the direction along which the object is moving; and hence $a = -9·8$ m s^{-2} in this case.

Suppose the ball was thrown straight up with an initial velocity, u, of 30 m s^{-1}. The time taken to reach the top of its motion can be obtained from the equation $v = u + at$. The velocity, v, at the top is zero; and since $u = 30$ m and $a = -9·8$ or 10 m s^{-2} (approx), we have

$$0 = 30 - 10t.$$

$$\therefore t = \frac{30}{10} = 3 \text{ s}.$$

The highest distance reached is thus given by

$$s = ut + \tfrac{1}{2}at^2$$
$$= 30 \times 3 - 5 \times 3^2 = 45 \text{ m}.$$

Resultant. Components

If a boy is running along the deck of a ship in a direction OA, and the ship is moving in a different direction OB, the boy will move relatively to the sea along a direction OC, between OA and OB, Fig. 1.6 (i). Now in one second the boat moves from O to B, where OB represents the velocity of the boat, a vector quantity, in magnitude and direction. The boy moves from O to A in the same time, where OA represents the velocity of the boy in magnitude and direction. Thus in one second the net effect relative to the sea is that the boy moves from O to C. It can now be seen that if lines OA, OB are drawn to represent in magnitude and direction the respective velocities of the boy and the ship, the magnitude and direction of the *resultant* velocity of the boy is represented by the diagonal OC of the completed parallelogram having OA, OB as two of its sides; OACB is known as a *parallelogram of velocities*. Conversely, a velocity represented completely by OC can be regarded as having an 'effective part', or *component* represented by OA, and another component represented by OB.

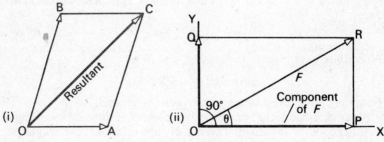

Fig. 1.6. Resultant and component.

In practice, we often require to find the component of a vector quantity in a certain direction. Suppose OR represents the vector F, and OX is the direction, Fig. 1.6 (ii). If we complete the parallelogram OQRP by drawing a perpendicular RP from R to OX, and a perpendicular RQ from R to OY, where OY is perpendicular to OX, we can see that OP, OQ represent the components of F along OX, OY respectively. Now the component OQ has no effect in a perpendicular direction; consequently OP represents the total effect of F along the direction OX. OP is called the 'resolved component' in this direction. If θ is the angle ROX, then, since triangle OPR has a right angle at P,

$$\text{OP} = \text{OR} \cos \theta = F \cos \theta \quad . \quad . \quad . \quad (4)$$

Components of g

The acceleration due to gravity, g, acts vertically downwards. In free fall, an object has an acceleration g. An object sliding freely down an inclined plane, however, has an acceleration due to gravity equal to the component of g down the plane. If it is inclined at 60° to the vertical, the acceleration down the plane is then $g \cos 60°$ or $9.8 \cos 60°$ m s^{-2}, which is 4.9 m s^{-2}.

Consider an object O thrown forward from the top of a cliff OA with a horizontal velocity u of 15 m s^{-1}. Fig. 1.7. Since u is horizontal, it has no component in a *vertical* direction. Similarly, since g acts vertically, it has no component in a *horizontal* direction.

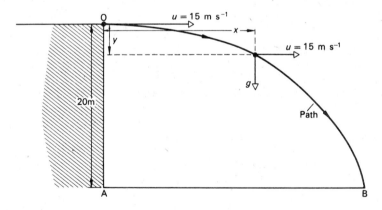

Fig. 1.7 Motion under gravity

We may thus treat vertical and horizontal motion independently. Consider the vertical motion from O. If OA is 20 m, the ball has an initial vertical velocity of zero and a vertical acceleration of g, which is

9.8 m s^{-2} (10 m s^{-2} approximately). Thus, from $s = ut + \frac{1}{2}at^2$, the time t to reach the bottom of the cliff is given, using $g = 10$ m s^{-2}, by

$$20 = \tfrac{1}{2} \cdot 10 \cdot t^2 = 5t^2, \text{ or } t = 2 \text{ s}.$$

So far as the horizontal motion is concerned, the ball continues to move forward with a constant velocity of 15 m s^{-1} since g has no component horizontally. In 2 seconds, therefore,

horizontal distance AB = distance from cliff = $15 \times 2 = 30$ m.

Generally, in a time t the ball falls a vertical distance, y say, from O given by $y = \frac{1}{2}gt^2$. In the same time the ball travels a horizontal distance, x say, from O given by $x = ut$, where u is the velocity of 15 m s^{-1}. If t is eliminated by using $t = x/u$ in $y = \frac{1}{2}gt^2$, we obtain $y = gx^2/2u$. This is the equation of a *parabola*. It is the path OB in Fig. 1.7.

Addition of Vectors

Suppose a ship is travelling due east at 30 km h^{-1} and a boy runs across the deck in a north-west direction at 6 km h^{-1}, Fig. 1.8 (i). We

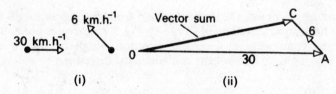

FIG. 1.8 Addition of vectors

can find the velocity and direction of the boy relative to the sea by adding the two velocities. Since velocity is a vector quantity, we draw a line OA to represent 30 km h^{-1} in magnitude and direction, and then, from the end of A, draw a line AC to represent 6 km h^{-1} in magnitude and direction, Fig. 1.8 (ii). The sum, or resultant, of the velocities is now represented by the line OC in magnitude and direction, because a distance moved in one second by the ship (represented by OA) together with a distance moved in one second by the boy (represented by AC) is equivalent to a movement of the boy from O to C relative to the sea.

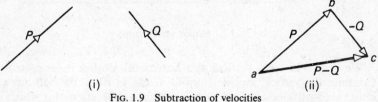

FIG. 1.9 Subtraction of velocities

In other words, the difference between the vectors $\vec{P}, \vec{Q}$ in Fig. 1.9 (i) is the *sum* of the vectors $\vec{P}$ and $(-\vec{Q})$. Now $(-\vec{Q})$ is a vector drawn exactly equal and opposite to the vector $\vec{Q}$. We therefore draw *ab* to represent $\vec{P}$ completely, and then draw *bc* to represent $(-\vec{Q})$ completely, Fig. 1.9 (ii). Then $\vec{P}+(-\vec{Q})$ = the vector represented by $ac = \vec{P}-\vec{Q}$.

Relative Velocity and Relative Acceleration

If a car A travelling at 50 km h^{-1} is moving in the same direction as another car B travelling at 60 km h^{-1}, the *relative velocity* of B to A = $60-50 = 10$ km h^{-1}. If, however, the cars are travelling in opposite directions, the relative velocity of B to A = $60-(-50) = 110$ km h^{-1}.

Suppose that a car X is travelling with a velocity v along a road 30° east of north, and a car Y is travelling with a velocity u along a road due east, Fig. 1.10 (i). Since 'velocity' has direction as well as magnitude, i.e., 'velocity' is a vector quantity (p. 2), we cannot subtract u and v numerically to find the relative velocity. We must adopt a method which takes into account the direction as well as the magnitude of the velocities, i.e., a vector subtraction is required.

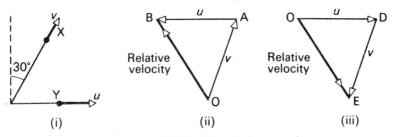

FIG. 1.10. Relative velocity.

The velocity of X relative to Y $= \vec{v}-u = \vec{v}+(-\vec{u})$. Suppose OA represents the velocity, v, of X in magnitude and direction, Fig. 1.10 (ii). Since Y is travelling due east, a velocity AB numerically equal to u but in the due *west* direction represents the vector $(-\vec{u})$. The vector sum of OA and AB is OB from p. 10, which therefore represents in magnitude and direction the velocity of X relative to Y. By drawing an accurate diagram of the two velocities, OB can be found.

The velocity of Y relative to X $= \vec{u}-\vec{v} = \vec{u}+(-\vec{v})$, and can be found by a similar method. In this case, OD represents the velocity, u, of Y in magnitude and direction, while DE represents the vector $(-\vec{v})$, which it is drawn numerically equal to v but in the *opposite* direction, Fig. 1.10 (iii). The vector sum of OD and DE is OE, which therefore represents the velocity of Y relative to X in magnitude and direction.

When two objects P, Q are each accelerating, the acceleration of P relative to Q = acceleration of P − acceleration of Q. Since 'acceleration' is a vector quantity, the relative acceleration must be found by vector subtraction, as for the case of relative velocity.

EXAMPLE

Explain the difference between a scalar and a vector quantity.
What is meant by the relative velocity of one body with respect to another? Two ships are 10 km apart on a line running S. to N. The one farther north is steaming W. at 20 km h^{-1}. The other is steaming N. at 20 km h^{-1}. What is their distance of closest approach and how long do they take to reach it? (*C.*)

Suppose the two ships are at X, Y, moving with velocities u, v respectively, each 20 km h^{-1} Fig. 1.11 (i). The velocity of Y relative to X $= \vec{v} - \vec{u} = \vec{v} + (-\vec{u})$. We therefore draw OA to represent $\vec{v}$ (20) and add to it AB, which represents $(-\vec{u})$, Fig. 1.11 (ii). The relative velocity is then represented by OB.

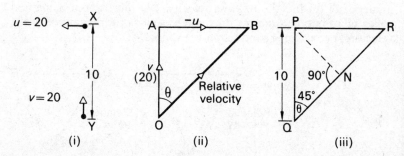

FIG. 1.11 Example

Since OAB is a right-angled triangle,

$$OB = \sqrt{OA^2 + AB^2} = \sqrt{20^2 + 20^2} = 28.28 \text{ km h}^{-1} \qquad \text{(i)}$$

Also, $\qquad \tan \theta = \dfrac{AB}{OA} = \dfrac{20}{20} = 1$, i.e., $\theta = 45°$ . . . . (ii)

Thus the ship Y will move along a direction QR relative to the ship X, where QR is at 45° to PQ, the north-south direction, Fig. 1.11 (iii). If PQ = 10 km, the distance of closest approach is PN, where PN is the perpendicular from P to QR.

$\therefore$ PN = PQ sin 45° = 10 sin 45° = 7.07 km.

The distance QN = 10 cos 45° = 7.07 km. Since, from (i), the relative velocity is 28.28 km h^{-1}, it follows that

$$\text{time to reach N} = \frac{7.07}{28.28} = \tfrac{1}{4} \text{ hour.}$$

LAWS OF MOTION. FORCE AND MOMENTUM

Newton's Laws of Motion

In 1687 SIR ISAAC NEWTON published a work called *Principia*, in which he expounded the Laws of Mechanics. He formulated in the book three 'laws of motion':

Law I. *Every body continues in its state of rest or uniform motion in a straight line, unless impressed forces act on it.*

Law II. *The change of momentum per unit time is proportional to the impressed force, and takes place in the direction of the straight line along which the force acts.*

Law III. *Action and reaction are always equal and opposite.*

These laws cannot be proved in a formal way; we believe they are correct because all the theoretical results obtained by assuming their truth agree with the experimental observations, as for example in astronomy (p. 66).

Inertia. Mass

Newton's first law expresses the idea of **inertia**. The inertia of a body is its reluctance to start moving, and its reluctance to stop once it has begun moving. Thus an object at rest begins to move only when it is pushed or pulled, i.e., when a *force* acts on it. An object O moving in a straight line with constant velocity will change its direction or move

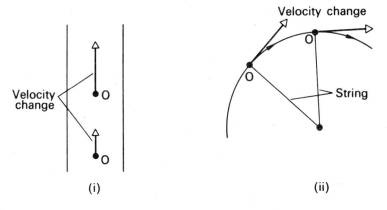

FIG. 1.12   Velocity changes

faster only if a new force acts on it. Fig. 1.12 (i). This can be demonstrated by a puck moving on a cushion of gas on a smooth level sheet of glass. As the puck slides over the glass, photographs taken at successive equal times by a stroboscopic method show that the motion is practically that of uniform velocity. Passengers in a bus or car are jerked forward when the vehicle stops suddenly. They continue in their state of motion until brought to rest by friction or collision. The use of safety belts reduces the shock.

Fig. 1.12 (ii) illustrates a velocity change when an object O is whirled at constant speed by a string. This time the magnitude of the velocity v is constant but its direction changes.

'Mass' is a measure of the inertia of a body. If an object changes its direction or its velocity slightly when a large force acts on it, its inertial mass is high. The mass of an object is constant all over the world; it is the same on the earth as on the moon. Mass is measured in kilogrammes (kg) or grammes (g) by means of a chemical balance, where it is compared with standard masses based on the International Prototype Kilogramme (see also p. 16).

Force. The newton

When an object X is moving it is said to have an amount of *momentum* given, by definition, by

$$\text{momentum} = \text{mass of } X \times \text{velocity} \qquad . \qquad . \qquad (1)$$

Thus an object of mass 20 kg moving with a velocity of 10 m s^{-1} has a momentum of 200 kg m s^{-1}. If another object collides with X its velocity alters, and thus the momentum of X alters. From Newton's second law, a *force* acts on X which is equal to the change in momentum per second.

Thus if F is the magnitude of a force acting on a constant mass m,

$$F \propto m \times \text{change of velocity per second}$$

$$\therefore F \propto ma,$$

where a is the *acceleration* produced by the force, by definition of a.

$$\therefore F = kma \qquad . \qquad . \qquad . \qquad . \qquad (2)$$

where k is a constant.

With SI units, the **newton** (N) is the unit of force. It is defined as the force which gives a mass of 1 kilogramme an acceleration of 1 metre s^{-2}. Substituting $F = 1\text{N}$, $m = 1$ kg and $a = 1$ m s^{-2} in the expression for F in (i), we obtain $k = 1$. Hence, with units as stated, $k = 1$.

$$\therefore \mathbf{F} = \mathbf{ma},$$

DYNAMICS

which is a standard equation in dynamics. Thus if a mass of 200 g is acted upon by a force F which produces an acceleration a of 4 m s^{-2}, then, since $m = 200 \text{ g} = 0.2 \text{ kg}$,

$$F = ma = 0.2(\text{kg}) \times 4(\text{m s}^{-2}) = 0.8 \text{ N}.$$

C.g.s. units of force

The *dyne* is the unit of force in the centimetre-gramme-second system; it is defined as the force acting on a mass of 1 gramme which gives it an acceleration of 1 cm s^{-2}. The equation $F = ma$ also applies when m is in gramme, a is in cm s^{-2}, and F is in dyne. Thus if a force of 10,000 dynes acts on a mass of 200 g, the acceleration a is given by

$$F = 10{,}000 = 200 \times a, \quad \text{or} \quad a = 50 \text{ cm s}^{-2}.$$

Suppose $m = 1 \text{ kg} = 1000 \text{ g}$, $a = 1 \text{ m s}^{-2} = 100 \text{ cm s}^{-2}$. Then, the force F is given by

$$F = ma = 1000 \times 100 \text{ dynes} = 10^5 \text{ dynes}.$$

But the force acting on a mass of 1 kg which gives it an acceleration of 1 m s^{-2} is the *newton*, N. Hence

$$1 \text{ N} = 10^5 \text{ dynes}$$

Weight. Relation between newton and kgf

The *weight* of an object is defined as the *force* acting on it due to gravity; the weight of an object can hence be measured by attaching it to a spring-balance and noting the extension, as the latter is proportional to the force acting on it (p. 57).

Suppose the weight of an object of mass m is denoted by W. If the object is released so that it falls to the ground, its acceleration is g. Now $F = ma$. Consequently the force acting on it, i.e., its weight, is given by

$$W = mg.$$

If the mass is 1 kg, then, since $g = 9.8 \text{ m s}^{-2}$, the weight $W = 1 \times 9.8 = 9.8 \text{ N}$ (newton). The force due to gravity on a mass of 1 kg where g is 9.80665 m s^{-2} is called a 1 *kilogramme force* or 1 kgf (1 kilogramme weight or 1 kg wt, however, varies with g). Hence it follows that

$$1 \text{ kgf} = 9.8 \text{ N} = 10 \text{ N approximately}.$$

A weight of 5 kgf is thus about 50 N. Further, $1 \text{ N} = \frac{1}{10} \text{ kgf approx} = 100 \text{ gf}$. It is interesting to note that the weight of an apple is about 1 newton.

We can see that on the surface of the earth, the value of g may be expressed as about '9·8 newton per kilogramme mass', or '9·8 N kg^{-1}'. The *force per unit mass* in a gravitational field is called the field *gravitational intensity*. On the moon's surface the gravitational intensity is about 1·6 N kg^{-1}.

The reader should note carefully the difference between the 'kilogramme' and the 'kilogramme force'; the former is a *mass* and is therefore constant all over the universe, whereas the kilogramme force is a *force* whose magnitude depends on the value of g. The acceleration due to gravity, g, depends on the distance of the place considered from the centre of the earth; it is slightly greater at the poles than at the equator, since the earth is not perfectly spherical (see p. 46). It therefore follows that the weight of an object differs in different parts of the world. On the moon, which is smaller than the earth and has a smaller density, an object would weigh about one-sixth of its weight on the earth.

The relation $F = ma$ can be verified by using a ticker-tape and timer to measure the acceleration of a moving trolley. Details are given in a more basic text, such as *Principles of Physics* (Chatto and Windus) by the author.

The following examples illustrate the application of $F = ma$. It should be carefully noted that (i) F represents the *resultant* force on the object of mass m, (ii) F must be expressed in the appropriate units of a 'force' and m in the corresponding units of a 'mass'.

EXAMPLES

1. A force of 200 N pulls a sledge of mass 50 kg and overcomes a constant frictional force of 40 N. What is the acceleration of the sledge?

$$\text{Resultant force, } F, = 200 - 40 = 160 \text{ N.}$$

From $F = ma$,

$$\therefore 160 = 50 \times a$$

$$\therefore a = 3.2 \text{ m s}^{-2}.$$

2. An object of mass 2·00 kg is attached to the hook of a spring-balance, and the latter is suspended vertically from the roof of a lift. What is the reading on the spring-balance when the lift is (i) ascending with an acceleration of 20 cm s^{-2}, (ii) descending with an acceleration of 10 cm s^{-2}, (iii) ascending with a uniform velocity of 15 cm s^{-1}. ($g = 10$ m s^{-2} or 10 N kg^{-1}.)

Suppose T is the tension (force) in the spring-balance in N.

(i) The object is acted upon two forces: (*a*) The tension T in the spring-balance, which acts upwards, (*b*) its weight, 20 N, which acts downwards. Since the object moves upwards, T is greater than 20 N. Hence the net force, F, acting on the object $= T - 20$ N. Now

$$F = ma,$$

where a is the acceleration in m s^{-2}, $0{\cdot}2$ m s^{-2}.

$$\therefore T - 20 \text{ N} = 2 \times a = 2 \times 0{\cdot}2 \text{ N}$$
$$\therefore T = 20{\cdot}4 \text{ N} \quad . \quad . \quad . \quad . \quad . \quad (1)$$

(ii) When the lift descends with an acceleration of 10 cm s^{-2} or $0{\cdot}1$ m s^{-2}, the weight, 20 N, is now greater than T_1, the tension in the spring-balance.

$$\therefore \text{resultant force} = 20 - T_1$$
$$\therefore F = 20 - T_1 = ma = 2 \times 0{\cdot}1$$
$$\therefore T_1 = 20 - 0{\cdot}2 = 19{\cdot}8 \text{ N}.$$

(iii) When the lift moves with constant velocity, the acceleration is zero. In this case the reading on the spring-balance is exactly equal to the weight, 20 N.

Linear Momentum

Newton defined the force acting on an object as the rate of change of its momentum, the momentum being the product of its mass and velocity (p. 13). *Momentum is thus a vector quantity.* Suppose that the mass of an object is m, its initial velocity is u, and its final velocity due to a force F acting on it for a time t is v. Then

$$\text{change of momentum} = mv - mu,$$

and hence
$$F = \frac{mv - mu}{t}$$
$$\therefore Ft = mv - mu = \text{momentum change} \quad . \quad . \quad (1)$$

The quantity Ft (force × time) is known as the *impulse* of the force on the object, and from (1) it follows that the units of momentum are the same as those of Ft, i.e., *newton second* (N s). From 'mass × velocity', alternative units are 'kg m s^{-1}'.

Force and momentum change

A person of weight 50 kgf, about 500 N, jumping from a height of 5 metres will land on the ground with a velocity $= \sqrt{2gh} = \sqrt{2 \times 10 \times 5} = 10$ m s^{-1}, assuming $g = 10$ m s^{-2} approx. If he does not flex his knees on landing, he will be brought to rest very quickly, say in $\frac{1}{10}$th second. The force F acting is then given by

$$F = \frac{\text{momentum change}}{\text{time}}$$
$$= \frac{50 \times 10}{\frac{1}{10}} = 5000 \text{ N}$$

This is a force of about 10 times the person's weight and this large force has a severe effect on the body.

Suppose, however, that the person flexes his knees and is brought to rest much more slowly on landing, say in 1 second. Then, from above, the force F now acting is 10 times less than before, or 500 N (approx). Consequently, much less damage is done to the person on landing.

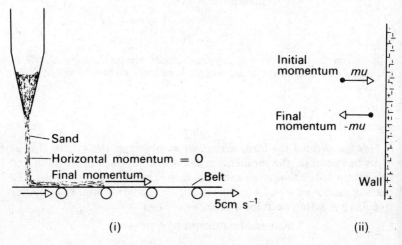

Fig. 1.13 Linear momentum

Suppose sand is allowed to fall vertically at a steady rate of 100 g s^{-1} on to a horizontal conveyor belt moving at a steady velocity of 5 cm s^{-1}. Fig. 1.13 (i). The initial horizontal velocity of the sand is zero. The final horizontal velocity is 5 cm s^{-1}. Now

mass = 100 g = 0·1 kg, velocity = 5 cm s^{-1} = 5×10^{-2} m s^{-1}

∴ momentum change per second = $0·1 \times 5 \times 10^{-2} = 5 \times 10^{-3}$ newton

= force on belt

Observe that this is a case where the *mass* changes with time and the velocity gained is constant. In terms of the calculus, the force is the rate of change of momentum mv, which is $v \times dm/dt$, and dm/dt is 100 g s^{-1} in this numerical example.

Consider a molecule of mass m in a gas, which strikes the wall of a vessel repeatedly with a velocity u and rebounds with a velocity $-u$. Fig. 1.13 (ii). Since momentum is a vector quantity, the momentum change = final momentum − initial momentum = $mu - (-mu) = 2mu$. If the containing vessel is a cube of side l, the molecule repeatedly takes a time $2l/u$ to make an impact with the same side.

DYNAMICS   19

∴ average force on wall due to molecule

$$= \frac{\text{momentum change}}{\text{time}}$$

$$= \frac{2mu}{2l/u} = \frac{mu^2}{l}.$$

The total gas pressure is the average force per unit area on the walls of the container due to all the numerous gas molecules.

EXAMPLES

1. A hose ejects water at a speed of 20 cm s^{-1} through a hole of area 100 cm^2. If the water strikes a wall normally, calculate the force on the wall in newton, assuming the velocity of the water normal to the wall is zero after collision.

The volume of water per second striking the wall $= 100 \times 20 = 2000$ cm^3.

∴ mass per second striking wall $= 2000$ g s$^{-1} = 2$ kg s^{-1}.

Velocity change of water on striking wall $= 20 - 0 = 20$ cm s$^{-1} = 0.2$ m s^{-1}.

∴ momentum change per second $= 2$ (kg s^{-1}) $\times 0.2$ (m s^{-1}) $= 0.4$ N.

2. Sand drops vertically at the rate of 2 kg s^{-1} on to a conveyor belt moving horizontally with a velocity of 0·1 m s^{-1}. Calculate (i) the extra power needed to keep the belt moving, (ii) the rate of change of kinetic energy of the sand. Why is the power twice as great as the rate of change of kinetic energy?

(i) Force required to keep belt moving = rate of increase of horizontal momentum of sand = mass per second $(dm/dt) \times$ velocity change $= 2 \times 0.1 = 0.2$ newton.

∴ power = work done per second = force × rate of displacement

= force × velocity $= 0.2 \times 0.1 = 0.02$ watt.

(ii) Kinetic energy of sand $= \frac{1}{2}mv^2$.

∴ rate of change of energy $= \frac{1}{2}v^2 \times \frac{dm}{dt}$, since v is constant,

$= \frac{1}{2} \times 0.1^2 \times 2 = 0.01$ watt.

Thus the power supplied is twice as great as the rate of change of kinetic energy. The extra power is due to the fact that the sand does not immediately assume the velocity of the belt, so that the belt at first moves relative to the sand. The extra power is needed to overcome the friction between the sand and belt.

Conservation of Linear Momentum

We now consider what happens to the linear momentum of objects which *collide* with each other.

Experimentally, this can be investigated by several methods:

1. Trolleys in collision, with ticker-tapes attached to measure velocities.
2. Linear Air-track, using perspex models in collision and stroboscopic photography for measuring velocities.

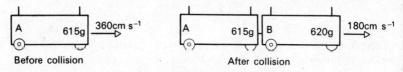

Before collision After collision

FIG. 1.14 Linear momentum experiment

As an illustration of the experimental results, the following measurements were taken in trolley collisions (Fig. 1.14):

Before collision.

Mass of trolley A = 615 g; initial velocity = 360 cm s^{-1}.

After collision.

A and B coalesced and both moved with velocity of 180 cm s^{-1}.
Thus the total linear momentum of A and B before collision = 0·615 (kg) × 3·6 (m s^{-1}) + 0 = 2·20 kg m s^{-1} (approx). The total momentum of A and B after collision = 1·235 × 1·8 = 2·20 kg m s^{-1} (approx).

Within the limits of experimental accuracy, it follows that *the total moment of A and B before collision = the total momentum after collision.* Similar results are obtained if A and B are moving with different speeds after collision, or in opposite directions before collision.

Principle of Conservation of Linear Momentum

These experimental results can be shown to follow from Newton's second and third laws of motion (p. 13).

Suppose that a moving object A, of mass m_1 and velocity u_1, collides with another object B, of mass m_2 and velocity u_2, moving in the same direction, Fig. 1.15. By Newton's law of action and reaction, the force F exerted by A on B is equal and opposite to that exerted by B on A. Moreover, the time t during which the force acted on B is equal to the time during which the force of reaction acted on A. Thus the magnitude of the impulse, Ft, on B is equal and opposite to the magnitude of the impulse on A. From equation (1), p. 17, the impulse is equal to the change of momentum. It therefore follows that the change in the total momentum of the two objects is *zero*, i.e., the total momentum of the

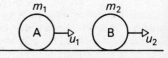

FIG. 1.15
Conservation of linear momentum

two objects is constant although a collision had occurred. Thus if A moves with a reduced velocity v_1 after collision, and B then moves with an increased velocity v_2,

$$m_1 u_1 + m_2 u_2 = m_1 v_1 + m_2 v_2.$$

The *principle of the conservation of linear momentum* states that, *if no external forces act on a system of colliding objects, the total momentum of the objects remains constant.*

EXAMPLES

1. An object A of mass 2 kg is moving with a velocity of 3 m s^{-1} and collides head on with an object B of mass 1 kg moving in the opposite direction with a velocity of 4 m s^{-1}. Fig. 1.16 (i). After collision both objects coalesce, so that they move with a common velocity v. Calculate v.

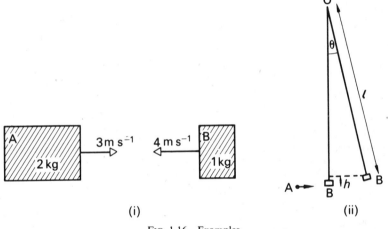

FIG. 1.16 Examples

Total momentum before collision of A and B in the direction of A

$$= 2 \times 3 - 1 \times 4 = 2 \text{ kg m s}^{-1}.$$

Note that momentum is a vector and the momentum of B is of opposite sign to A.

After collision, momentum of A and B in the direction of A $= 2v + 1v = 3v$.

$$\therefore 3v = 2$$

$$\therefore v = \tfrac{2}{3} \text{ m s}^{-1}.$$

2. What is understood by (*a*) the principle of the *conservation of energy*, (*b*) the principle of the *conservation of momentum*?

A bullet of mass 20 g travelling horizontally at 100 m s^{-1}, embeds itself in the centre of a block of wood of mass 1 kg which is suspended by light vertical strings 1 m in length. Calculate the maximum inclination of the strings to the vertical.

Describe in detail how the experiment might be carried out and used to determine the velocity of the bullet just before the impact of the block. (*N*.)

Second part. Suppose A is the bullet, B is the block suspended from a point O, and θ is the maximum inclination to the vertical, Fig. 1.16 (ii). If v cm s^{-1} is the common velocity of block and bullet when the latter is brought to rest relative to the block, then, from the principle of the conservation of momentum, since 20 g = 0·02 kg,

$$(1 + 0.02)v = 0.02 \times 100$$

$$\therefore v = \frac{2}{1 \cdot 02} = \frac{100}{51} \text{ m s}^{-1}.$$

The vertical height risen by block and bullet is given by $v^2 = 2gh$, where $g = 9.8$ m s^{-2} and $h = l - l\cos\theta = l(1 - \cos\theta)$.

$$\therefore v^2 = 2gl(1 - \cos\theta).$$

$$\therefore \left(\frac{100}{51}\right)^2 = 2 \times 9.8 \times 1(1 - \cos\theta).$$

$$\therefore 1 - \cos\theta = \left(\frac{100}{51}\right)^2 \times \frac{1}{2 \times 9.8} = 0.1962.$$

$$\therefore \cos\theta = 0.8038, \text{ or } \theta = 37° \text{ (approx.).}$$

The velocity, v, of the bullet can be determined by applying the conservation of momentum principle.

Thus $mv = (m + M)V$, where m is the mass of the bullet, M is the mass of the block, and V is the common velocity. Then $v = (m + M)V/m$. The quantities m and M can be found by weighing. V is calculated from the horizontal displacement a of the block, since (i) $V^2 = 2gh$ and (ii) $h(2l - h) = a^2$ from the geometry of the circle, so that, to a good approximation, $2h = a^2/l$.

Inelastic and elastic collisions

In collisions, the total momentum of the colliding objects is always conserved. Usually, however, their total kinetic energy is not conserved. Some of it is changed to heat or sound energy, which is not recoverable. Such collisions are said to be *inelastic*. If the total kinetic energy is conserved, the collision is said to be *elastic*. The collision between two smooth billiard balls is approximately elastic. Many atomic collisions are elastic. Electrons may make elastic or inelastic collisions with atoms of a gas. As proved on p. 28, the kinetic energy of a mass m moving with a velocity v has kinetic energy equal to $\frac{1}{2}mv^2$.

As an illustration of the mechanics associated with elastic collisions, consider a sphere A of mass m and velocity v incident on a stationary sphere B of equal mass m. (Fig. 1.17 (i). Suppose the collision is elastic, and after collision let A move with a velocity v_1 at an angle of 60° to its original direction and B move with a velocity v_2 at an angle θ to the direction of v.

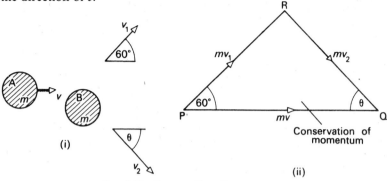

FIG. 1.17 Conservation of momentum

Since momentum is a vector (p. 17), we may represent the momentum mv of A by the line PQ drawn in the direction of v. Fig. 1.17 (ii). Likewise, PR represents the momentum mv_1 of A after collision. *Since momentum is conserved, the vector RQ must represent the momentum mv_2 of B after collision*, that is,

$$m\vec{v} = m\vec{v}_1 + m\vec{v}_2.$$

Hence
$$\vec{v} = \vec{v}_1 + \vec{v}_2,$$

or PQ represents v in magnitude, PR represents v_1 and RQ represents v_2. But if the collision is elastic,

$$\tfrac{1}{2}mv^2 = \tfrac{1}{2}mv_1^2 + \tfrac{1}{2}mv_2^2$$
$$\therefore v^2 = v_1^2 + v_2^2.$$

Consequently, triangle PRQ is a right-angled triangle with angle R equal to 90°.

$$\therefore v_1 = v \cos 60° = \frac{v}{2}.$$

Also, $\theta = 90° - 60° = 30°$, and $v_2 = v \cos 30° = \frac{\sqrt{3}v}{2}$.

Coefficient of restitution

In practice, colliding objects do not stick together and kinetic energy is always lost. If a ball X moving with velocity u_1 collides head-on with a ball Y moving with a velocity u_2 in the same direction, then Y will move faster with a velocity v_1 say and X may then have a reduced velocity v_2 in the same direction. The coefficient of restitution, e, between X and Y is defined as the ratio:

$$\frac{\text{velocity of separation}}{\text{velocity of approach}} \quad \text{or} \quad \frac{v_2 - v_1}{u_1 - u_2}.$$

The coefficient of restitution is approximately constant between two given materials. It varies from $e = 0$, when objects stick together and the collision is completely inelastic, to $e = 1$, when objects are very hard and the collision is practically elastic. Thus, from above, if $u_1 = 4$ m s^{-1}, $u_2 = 1$ m s^{-1} and $e = 0.8$, then velocity of separation, $v_2 - v_1 = 0.8 \times (4-1) = 2.4$ m s^{-1}.

Momentum and Explosive forces

There are numerous cases where momentum changes are produced by *explosive* forces. An example is a bullet of mass $m = 50$ g say, fired from a rifle of mass $M = 2$ kg with a velocity v of 100 m s^{-1}. Initially, the total momentum of the bullet and rifle is zero. From the principle of the conservation of linear momentum, when the bullet is fired the total momentum of bullet and rifle is still zero, since no external force has acted on them. Thus if V is the velocity of the rifle,

$$mv \text{ (bullet)} + MV \text{(rifle)} = 0$$

$$\therefore MV = -mv, \quad \text{or} \quad V = -\frac{m}{M}v.$$

The momentum of the rifle is thus *equal and opposite* to that of the bullet. Further, $V/v = -m/M$. Since $m/M = 50/2000 = 1/40$, it follows that $V = -v/40 = 2.5$ m s^{-1}. This means that the rifle moves back or *recoils* with a velocity only about $\frac{1}{40}$th that of the bullet.

If it is preferred, one may also say that the explosive force produces the same numerical momentum change in the bullet as in the rifle. Thus $mv = MV$, where V is the velocity of the rifle in the *opposite* direction to that of the bullet. From p. 28,
 the kinetic energy, E_1, of the bullet $= \frac{1}{2}mv^2 = \frac{1}{2} \cdot 0.05 \cdot 100^2 = 250$J
 the kinetic energy, E_2, of the rifle $= \frac{1}{2}MV^2 = \frac{1}{2} \cdot 2 \cdot 2.5^2 = 6.25$J
Thus the total kinetic energy produced by the explosion = 256.25 J. The kinetic energy E_1 of the bullet is thus 250/256.25, or about 98%, of the total energy. This is explained by the fact that the kinetic energy depends on the *square* of the velocity. The high velocity of the bullet thus more than compensates for its small mass relative to that of the rifle. See also p. 28.

Rocket

Consider a rocket moving in outer space where no external forces act on it. Suppose its mass is M and its velocity is v at a particular instant. Fig. 1.18 (i). When a mass m of fuel is ejected, the mass of the rocket becomes $(M-m)$ and its velocity increases to $(v+\Delta v)$. Fig. 1.18 (ii).

Suppose the fuel is always ejected at a constant speed u relative to

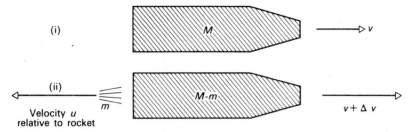

Fig. 1.18. Motion of rocket

the rocket. Then the velocity of the mass $m = v + \dfrac{\Delta v}{2} - u$ in the direction of the rocket, since the initial velocity of the rocket is v and the final velocity is $v + \Delta v$, an average of $v + \Delta v/2$.

We now apply the principle of the conservation of momentum to the rocket and fuel. *Initially*, before m of fuel was ejected, momentum of rocket and fuel inside rocket $= Mv$.

After m is ejected, momentum of rocket $= (M-m)(v+\Delta v)$

and momentum of fuel $= m\left(v + \dfrac{\Delta v}{2} - u\right)$.

$$\therefore (M-m)(v+\Delta v) + m\left(v + \dfrac{\Delta v}{2} - u\right) = Mv.$$

Neglecting the product of $m \cdot \Delta v$, then, after simplification,

$$M \cdot \Delta v - mu = 0,$$

$$\therefore \frac{m}{M} = \frac{\Delta v}{u}.$$

Now $\qquad m = $ mass of fuel ejected $= -\Delta M$,

$$\therefore -\frac{\Delta M}{M} = \frac{\Delta v}{u}.$$

Integrating between limits of M, M_0 and v, v_0 respectively

$$\therefore \int_{M_0}^{M} -\frac{\Delta M}{M} = \frac{1}{u}\int_{v_0}^{v} \Delta v.$$

$$\therefore -\log_e \frac{M}{M_0} = \frac{v - v_0}{u}.$$

$$\therefore M = M_0 e^{-(v-v_0)/u} \qquad . \qquad . \qquad . \qquad (1)$$

or $\qquad v = v_0 - u \log_e(M/M_0) \qquad . \qquad . \qquad . \qquad (2)$

When the mass M decreases to $M_0/2$

$$v = v_0 + u \log_e 2.$$

Motion of centre of mass

If two particles, masses m_1 and m_2, are distances x_1, x_2 respectively from a given axis, their *centre of mass* is at a distance x from the axis given by $m_1 x_1 + m_2 x_2 = (m_1 + m_2)\bar{x}$. See p. 118. Since velocity, $v = dx/dt$ generally, the velocity $\bar{v}$ of the centre of mass in the particular direction is given by $m_1 v_1 + m_2 v_2 = (m_1 + m_2)\bar{v}$, where v_1, v_2 are the respective velocities of m_1, m_2. The quantity $(m_1 v_1 + m_2 v_2)$ represents the total momentum of the two particles. The quantity $(m_1 + m_2)\bar{v} = M\bar{v}$, where M is the total mass of the particles. Thus we can imagine that the total mass of the particles is concentrated at the centre of mass while they move, and that the velocity $\bar{v}$ of the centre of mass is always given by *total momentum* $= M\bar{v}$.

If *internal forces* act on the particles while moving, then, since action and reaction are equal and opposite, their resultant on the whole body is zero. Consequently the total momentum is unchanged and hence the velocity or motion of their centre of mass is unaffected. If an *external force*, however, acts on the particles, the total momentum is changed. The motion of their centre of mass now follows a path which is due to the external force.

We can apply this to the case of a shell fired from a gun. The centre of mass of the shell follows at first a parabolic path. This is due to the external force of gravity, its weight. If the shell explodes in mid-air, the fragments fly off in different directions. But the numerous internal forces which occur in the explosion have zero resultant, since action and reaction are equal and opposite and the forces can all be paired. Consequently *the centre of mass of all the fragments continues to follow the same parabolic path*. As soon as one fragment reaches the ground, an external force now acts on the system of particles. A different parabolic path is then followed by the centre of mass.

If a bullet is fired in a horizontal direction from a rifle, where is their centre of mass while the bullet and rifle are both moving?

Work

When an engine pulls a train with a constant force of 50 units through a distance of 20 units in its own direction, the engine is said by definition to do an amount of *work* equal to 50×20 or 1,000 units, the product of the force and the distance. Thus if W is the amount of work,
$W = $ force $\times$ distance moved in direction of force.

Work is a *scalar* quantity; it has no property of direction but only magnitude. When the force is one newton and the distance moved is one metre, then the work done is one *joule*. Thus a force of 50 N moving through a distance of 10 m does 50×10 or 500 joule of work. Note this is also a measure of the *energy* transferred to the object.

DYNAMICS

The force to raise steadily a mass of 1 kg is equal to its weight, which is about 10 N (see p. 15). Thus if the mass of 1 kg is raised vertically through 1 m, then, approximately, work done = 10 (N) × 1 (m) = 10 joule.

The c.g.s. unit of work is the *erg*; it is the work done when a force of 1 dyne moves through 1 cm. Since 1 N = 10^5 dynes and 1 m = 100 cm, then 1 N moving through 1 m does an amount of work = 10^5 (dyne) × 100 (cm) = 10^7 ergs = 1 joule, by definition of the joule (p. 26).

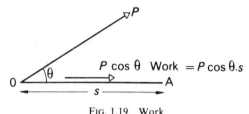

FIG. 1.19 Work

Before leaving the topic of 'work', the reader should note carefully that we have assumed the force to move an object in its own direction. Suppose, however, that a force P pulls an object a distance s along a line OA acting at an angle θ to it, Fig. 1.19. The component of P along OA is $P \cos \theta$ (p. 8), and this is the effective part of P pulling along the direction OA. The component of P along a direction perpendicular to OA has no effect along OA. Consequently

$$\text{work done} = P \cos \theta \times s.$$

In general, the work done by a force is equal to the product of the force and the displacement of its point of application in the direction of the force.

Power

When an engine does work quickly, it is said to be operating at a high *power*; if it does work slowly it is said to be operating at a low power. 'Power' is defined as the *work done per second*, i.e.,

$$\text{power} = \frac{\text{work done}}{\text{time taken}}.$$

The practical unit of power, the SI unit, is 'joule per second' or *watt* (W); the watt is defined as the rate of working at 1 joule per second.

$$1 \text{ horse-power (hp)} = 746 \text{ W} = \tfrac{3}{4} \text{ kW (approx)},$$

where 1 kW = 1 kilowatt or 1000 watt. Thus a small motor of $\frac{1}{6}$ hp in a vacuum carpet cleaner has a power of about 125 W.

Kinetic Energy

An object is said to possess *energy* if it can do work. When an object possesses energy because it is moving, the energy is said to be *kinetic*, e.g., a flying stone can disrupt a window. Suppose that an object of mass m is moving with a velocity u, and is gradually brought to rest in a distance s by a constant force F acting against it. The kinetic energy originally possessed by the object is equal to the work done against F, and hence

$$\text{kinetic energy} = F \times s.$$

But $F = ma$, where a is the retardation of the object. Hence $F \times s = mas$. From $v^2 = u^2 + 2as$ (see p. 6), we have, since $v = 0$ and a is negative in this case,

$$0 = u^2 - 2as, \text{ i.e., } as = \frac{u^2}{2}.$$

$\therefore$ *kinetic energy* $= mas = \frac{1}{2}mu^2$.

When m is in kg and u is in m s^{-1}, then $\frac{1}{2}mu^2$ is in *joule*. Thus a car of mass 1000 kg, moving with a velocity of 36 km h^{-1} or 10 m s^{-1}, has an amount W of kinetic energy given by

$$W = \tfrac{1}{2}mu^2 = \tfrac{1}{2} \times 1000 \times 10^2 = 50{,}000 \text{ J}$$

Kinetic Energies due to Explosive Forces

Suppose that, due to an explosion or nuclear reaction, a particle of mass m breaks away from the total mass concerned and moves with velocity v, and a mass M is left which moves with velocity V in the opposite direction. Then

$$\frac{\text{kinetic energy, } E_1, \text{ of mass } m}{\text{kinetic energy, } E_2, \text{ of mass } M} = \frac{\tfrac{1}{2}mv^2}{\tfrac{1}{2}MV^2} = \frac{mv^2}{MV^2} \qquad (1)$$

Now from the principle of the conservation of linear momentum, $mv = MV$. Thus $v = MV/m$. Substituting for v in (1).

$$\therefore \frac{E_1}{E_2} = \frac{mM^2V^2}{m^2MV^2} = \frac{M}{m} = \frac{1/m}{1/M}.$$

Hence the energy is *inversely*-proportional to the masses of the particles, that is, the smaller mass, m say, has the larger energy. Thus if E is the total energy of the two masses, the energy of the smaller mass $= ME/(M+m)$. An α-particle has a mass of 4 units and a radium nucleus a mass of 228 units. If disintegration of a thorium nucleus,

DYNAMICS 29

mass 232, produces an α-particle and radium nucleus, and a release of energy of 4·05 MeV, where 1 MeV = $1·6 \times 10^{-13}$ J, then

$$\text{energy of } \alpha\text{-particle} = \frac{228}{(4+228)} \times 4·05 = 3·98 \text{ MeV}.$$

The α-particle thus travels a relatively long distance before coming to rest compared to the radium nucleus.

Potential Energy

A weight held stationary above the ground has energy, because, when released, it can raise another object attached to it by a rope passing over a pulley, for example. A coiled spring also has energy, which is released gradually as the spring uncoils. The energy of the weight or spring is called *potential energy*, because it arises from the position or arrangement of the body and not from its motion. In the case of the weight, the energy given to it is equal to the work done by the person or machine which raises it steadily to that position against the force of attraction of the earth. In the case of the spring, the energy is equal to the work done in displacing the molecules from their normal equilibrium positions against the forces of attraction of the surrounding molecules.

If the mass of an object is m, and the object is held stationary at a height h above the ground, the energy released when the object falls to the ground is equal to the work done

$$= \text{force} \times \text{distance} = \text{weight of object} \times h.$$

Suppose the mass m is 5 kg, so that the weight is $5 \times 9·8$ N or 50 N approx, and h is 4 metre. Then

$$\text{potential energy P.E.} = 50 \text{ (N)} \times 4 \text{ (m)} = 200 \text{ J}$$

(more accurately, P.E. = 196 J).

Generally, at a height of h,

$$\text{potential energy} = mgh,$$

where m is in kg, h is in metre, g = 9·8.

EXAMPLE

Define *work, kinetic energy, potential energy*. Give *one* example of *each* of the following: (*a*) the conversion into kinetic energy of the work done on a body and (*b*) the conversion into potential energy of the work done on a body.

A rectangular block of mass 10 g rests on a rough plane which is inclined to the horizontal at an angle $\sin^{-1}$ (0·05). A force of 0·03 newton, acting in a direction parallel to a line of greatest slope, is applied to the block so that it moves up the plane. When the block has travelled a distance of 110 cm from its initial position, the applied force is removed. The block moves on and comes to rest again after travelling a further 25 cm. Calculate (i) the work done by the applied force, (ii) the gain in potential energy of the block and (iii) the value of the coefficient of sliding friction between the block and the surface of the inclined plane. How would the coefficient of sliding friction be measured if the angle of the slope could be altered? (*O. and C.*)

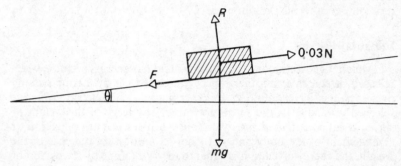

FIG. 1.20 Example

(i) Force = 0·03 newton; distance = 110 cm = 1·1 m.

$$\therefore \text{work} = 0{\cdot}03 \times 1{\cdot}1 = 0{\cdot}033 \text{ J}.$$

(ii) Gain in P.E. = wt × height moved = $0{\cdot}01 \times 9{\cdot}8$ N × $1{\cdot}35 \sin\theta$ m,

$$= 0{\cdot}01 \times 9{\cdot}8 \text{ N} \times 1{\cdot}35 \times 0{\cdot}05 \text{ m} = 0{\cdot}0066 \text{ J (approx.)}.$$

(iii) Work done against frictional force F = work done by force − gain in P.E.

$$= 0{\cdot}033 - 0{\cdot}0066 = 0{\cdot}0264 \text{ J}.$$

$$\therefore F \times 1{\cdot}35 = 0{\cdot}0264.$$

$$\therefore F = \frac{0{\cdot}0264}{1{\cdot}35} \text{ N}.$$

Normal reaction, $R = mg \cos\theta = mg$ (approx.), since θ is so small

$$\therefore \mu = \frac{F}{R} = \frac{0{\cdot}0264}{1{\cdot}35 \times 0{\cdot}01 \times 9{\cdot}8} = 0{\cdot}2 \text{ (approx.)}.$$

Conservative Forces

If a ball of weight W is raised steadily from the ground to a point X at a height h above the ground, the work done is $W.h$. The potential energy, P.E., of the ball is thus $W.h$. Now whatever route is taken from ground level to X, the work done is the same—if a *longer* path is chosen, for example, the component of the weight in the particular direction must then be overcome and so the force required to move the ball is correspondingly smaller. The P.E. of the ball at X is thus independent of the route to X. This implies that if the ball is taken in a closed path round to X again, *the total work done is zero*. Work has been expended on one part of the closed path, and regained on the remaining part.

When the work done in moving round a closed path in a field to the original point is zero, the forces in the field are called *conservative forces*. The earth's gravitational field is an example of a field containing conservative forces, as we now show.

Suppose the ball falls from a place Y at a height h to another X at a height of x above the ground. Fig. 1.21. Then, if W is the weight of the ball and m its mass,

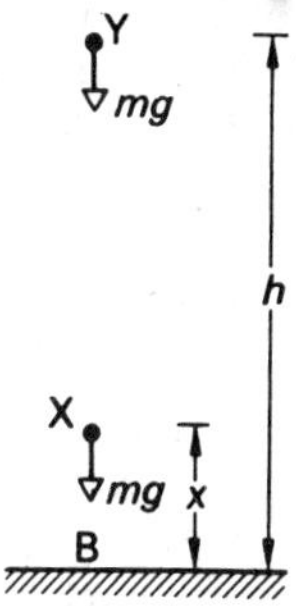

P.E. at X $= Wx = mgx$

and K.E. at X $= \frac{1}{2}mv^2 = \frac{1}{2}m \cdot 2g(h-x) = mg(h-x)$,

using $v^2 = 2as = 2g(h - x)$. Hence

P.E. $+$ K.E. $= mgx + mg(h-x) = mgh$.

Fig. 1.21. Mechanical energy

Thus at any point such as X, the total mechanical energy of the falling ball is equal to the original energy. The mechanical energy is hence constant or conserved. This is the case for a conservative field.

Non-Conservative forces. Principle of Conservation of Energy

The work done in taking a mass m round a closed path in the conservative earth's gravitational field is zero. Fig. 1.22 (i). If the work done in taking an object round a closed path to its original position is

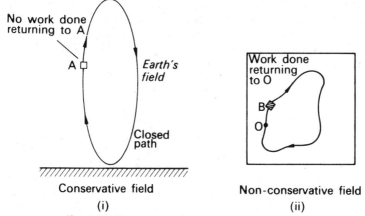

Fig. 1.22 Non-conservative and conservative fields

not zero, the forces in the field are said to be *non-conservative*. This is the case, for example, when a wooden block B is pushed round a closed path on a rough table to its initial position O. Work is therefore done against friction, both as A moves away from O and as it returns. In a conservative field, however, work is done during part of the path and regained for the remaining part.

When a body falls in the earth's gravitational field, a small part of the energy is used up in overcoming the resistance of the air. This energy is dissipated or lost as heat—it is not regained in moving the body back to its original position. This resistance is another example of the action of a non-conservative force.

Although energy may be transformed from one form to another, as in the last example from mechanical energy to heat, *the total energy in a given system is always constant*. If an electric motor is supplied with 1000 joule of energy, 850 joule of mechanical energy, 140 joule of heat energy and 10 joule of sound energy may be produced. This is called the *Principle of the Conservation of Energy* and is one of the key principles in science.

Mass and Energy

Newton said that the 'mass' of an object was 'a measure of the quantity of matter' in it. In 1905, Einstein proved from his Special Theory of Relativity that energy is released from an object when its mass decreases. His mass-energy relation states that if the mass decreases by Δm kg, the energy released in joule, ΔW, is given by

$$\Delta W = \Delta m \cdot c^2,$$

where c is the numerical value of the speed of light in m s^{-1}, which is 3×10^8. Experiments in Radioactivity on nuclear reactions showed that Einstein's relation was true. Thus mass is a form of energy.

Einstein's relation shows that even if a small change in mass occurs, a relatively large amount of energy is produced. Thus if $\Delta m = 1$ milligramme $= 10^{-6}$ kg, the energy ΔW released

$$= \Delta m \cdot c^2 = 10^{-6} \times (3 \times 10^8)^2 = 9 \times 10^{10} \text{ J}.$$

This energy will keep 250,000 100-W lamps burning for about an hour. In practice, significant mass changes occur only in nuclear reactions.

The internal energy of a body of mass m may be considered as $E_{int} = mc^2$, where m is its rest mass. In nuclear reactions where two particles collide, a change occurs in their total kinetic energy and in their total mass. The increase in total kinetic energy is accompanied by an equal decrease in internal energy, $\Delta m \cdot c^2$. Thus the total energy, kinetic plus internal, remains constant.

Before Einstein's mass-energy relation was known, two independent laws of science were:

(1) *The Principle of the Conservation of Mass* (the total mass of a given system of objects is constant even though collisions or other actions took place between them);

(2) *The Principle of the Conservation of Energy* (the total energy of a given system is constant). From Einstein's relation, however, the two laws can be combined into one, namely, the Principle of the Conservation of Energy.

The summary below may assist the reader; it refers to the units of some of the quantities encountered, and their relations.

Quantity	SI	C.G.S.	Relations
Force (vector)	newton (N)	dyne	10^5 dyne = 1 N 1 kgf = 9·8 N (approx, 10 N) 1 gf = 0·0098 N (approx, 0·01 N)
Mass (scalar)	kilogramme (kg)	gramme (g)	
Momentum (vector)	newton second (Ns)	dyne second	
Energy (scalar)	joule (J)	erg	10^7 erg = 1 J
Power (scalar)	watt (W)	erg s^{-1}	1 W = 1 J s^{-1} 1 h.p. = 746 W

Dimensions

By the *dimensions* of a physical quantity we mean the way it is related to the fundamental quantities mass, length and time; these are usually denoted by M, L, and T respectively. An area, length × breadth, has dimensions L × L or L^2; a volume has dimensions L^3; density, which is mass/volume, has dimensions M/L^3 or ML^{-3}; relative density has no dimensions, since it is the ratio of similar quantities, in this case two masses (p. 132); an angle has no dimensions, since it is the ratio of two lengths.

As an area has dimensions L^2, the *unit* may be written in terms of the metre as 'm^2'. Similarly, the dimensions of a volume are L^3 and hence the unit is 'm^3'. Density has dimensions ML^{-3}. The density of mercury is thus written as '13 600 kg m^{-3}'. If some physical quantity has dimensions $ML^{-1}T^{-1}$, its unit may be written as 'kg m^{-1} s^{-1}'.

The following are the dimensions of some quantities in Mechanics:

Velocity. Since velocity = $\dfrac{\text{distance}}{\text{time}}$, its dimensions are L/T or LT^{-1}.

Acceleration. The dimensions are those of velocity/time, i.e., L/T^2 or LT^{-2}.

Force. Since force = mass × acceleration, its dimensions are MLT^{-2}.

Work or Energy. Since work = force × distance, its dimensions are ML^2T^{-2}.

EXAMPLE

In the gas equation $(p + \frac{a}{V^2})(V-b) = RT$, what are the dimensions of the constants a and b?

p represents pressure, V represents volume. The quantity a/V^2 must represent a pressure since it is added to p. The dimensions of p = [force]/[area] = $MLT^{-2}/L^2 = ML^{-1}T^{-2}$; the dimensions of $V = L^3$. Hence

$$\frac{[a]}{L^6} = ML^{-1}T^{-2}, \text{ or } [a] = ML^5T^{-2}.$$

The constant b must represent a volume since it is subtracted from V. Hence

$$[b] = L^3.$$

Application of Dimensions. Simple Pendulum

If a small mass is suspended from a long thread so as to form a simple pendulum, we may reasonably suppose that the period, T, of the oscillations depends only on the mass m, the length l of the thread, and the acceleration, g, due to gravity at the place concerned. Suppose then that

$$T = km^x l^y g^z \qquad \qquad \qquad \text{(i)}$$

where x, y, z, k are unknown numbers. The dimensions of g are LT^{-2} from above. Now the dimensions of both sides of (i) must be the same.

$$\therefore T = M^x L^y (LT^{-2})^z.$$

Equating the indices of M, L, T on both sides, we have

$$x = 0,$$
$$y + z = 0,$$

and
$$-2z = 1.$$
$$\therefore z = -\tfrac{1}{2}, y = \tfrac{1}{2}, x = 0.$$

Thus, from (i), the period T is given by

$$T = kl^{\frac{1}{2}} g^{-\frac{1}{2}},$$

or
$$T = k\sqrt{\frac{l}{g}}.$$

DYNAMICS

We cannot find the magnitude of k by the method of dimensions, since it is a number. A complete mathematical investigation shows that $k = 2\pi$ in this case, and hence $T = 2\pi\sqrt{l/g}$. (See also p. 55).

Velocity of Transverse Wave in a String

As another illustration of the use of dimensions, consider a wave set up in a stretched string by plucking it. The velocity, V, of the wave depends on the tension, F, in the string, its length l, and its mass m, and we can therefore suppose that

$$V = kF^x l^y m^z, \qquad \qquad \text{(i)}$$

where x, y, z are numbers we hope to find by dimensions and k is a constant.

The dimensions of velocity, V, are LT^{-1}, the dimensions of tension, F, are MLT^{-2}, the dimension of length, l, is L, and the dimension of mass, m, is M. From (i), it follows that

$$LT^{-1} \equiv (MLT^{-2})^x \times L^y \times M^z.$$

Equating powers of M, L, and T on both sides,

$$\therefore 0 = x+z, \qquad \qquad \text{(i)}$$
$$1 = x+y, \qquad \qquad \text{(ii)}$$
and $$-1 = -2x, \qquad \qquad \text{(iii)}$$

$$\therefore x = \tfrac{1}{2}, z = -\tfrac{1}{2}, y = \tfrac{1}{2}.$$
$$\therefore V = k \cdot F^{\frac{1}{2}} l^{\frac{1}{2}} m^{-\frac{1}{2}},$$
$$\text{or } V = k\sqrt{\frac{Fl}{m}} = k\sqrt{\frac{F}{m/l}} = k\sqrt{\frac{\text{Tension}}{\text{mass per unit length}}}$$

A complete mathematical investigation shows that $k = 1$.

The method of dimensions can thus be used to find the relation between quantities when the mathematics is too difficult. It has been extensively used in hydrodynamics, for example. See also pp. 205, 212.

EXERCISES 1

(*Assume $g = 10\ m\ s^{-2}$, unless otherwise given*)

What are the missing words in the statements 1–10?

1. The dimensions of velocity are ...

2. The dimensions of force are ...

3. Using 'vector' or 'scalar', (i) mass is a ... (ii) force is a ... (iii) energy is a ... (iv) momentum is a ...

36   MECHANICS AND PROPERTIES OF MATTER

4. Linear momentum is defined as . . .
5. An 'elastic' collision is one in which the . . . and the . . . are conserved.
6. When two objects collide, their . . . is constant provided no . . . forces act.
7. One newton × one metre = . . .
8. 1 kilogram force (1 kilogram wt) = . . . newton, approx.
9. The momentum of two different bodies must be added by a . . . method.
10. Force is the . . . of change of momentum.

Which of the following answers, A, B, C, D or E, do you consider is the correct one in the statements 11–14?

11. When water from a hosepipe is incident horizontally on a wall, the force on the wall is calculated from A speed of water, B mass × velocity, C mass per second × velocity, D energy of water, E momentum change.

12. When a ball of mass 2 kg moving with a velocity of 10 m s^{-1} collides head-on with a ball of mass 3 kg and both move together after collision, the common velocity is A 5 m s^{-1} and energy is lost, B 4 m s^{-1} and energy is lost, C 2 m s^{-1} and energy is gained, D 6 m s^{-1} and momentum is gained, E 6 m s^{-1} and energy is conserved.

13. An object of mass 2 kg moving with a velocity of 4 m s^{-1} has a kinetic energy of A 8 joule, B 16 erg, C 4000 erg, D 16 joule, E 40 000 joule.

14. The dimensions of work are A ML^2T^{-2} and it is a scalar, B ML^2T^{-2} and it is a vector, C MLT^{-1} and it is a scalar, D ML^2T and it is a scalar, E MLT and it is a vector.

15. A car moving with a velocity of 36 km h^{-1} accelerates uniformly at 1 m s^{-2} until it reaches a velocity of 54 km h^{-1}. Calculate (i) the time taken, (ii) the distance travelled during the acceleration, (iii) the velocity reached 100 m from the place where the acceleration began.

16. A ball of mass 100 g is thrown vertically upwards with an initial speed of 72 km h^{-1}. Calculate (i) the time taken to return to the thrower, (ii) the maximum height reached, (iii) the kinetic and potential energies of the ball half-way up.

17. The velocity of a ship A relative to a ship B is 10·0 km h^{-1} in a direction N. 45° E. If the velocity of B is 20·0 km h^{-1} in a direction N. 60° W., find the actual velocity of A in magnitude and direction.

18. Calculate the energy of (i) a 2 kg object moving with a velocity of 10 m s^{-1}, (ii) a 10 kg object held stationary 5 m above the ground.

19. A 4 kg ball moving with a velocity of 10·0 m s^{-1} collides with a 16 kg ball moving with a velocity of 4·0 m s^{-1} (i) in the same direction, (ii) in the opposite direction. Calculate the velocity of the balls in each case if they coalesce on impact, and the loss of energy resulting from the impact. State the principle used to calculate the velocity.

DYNAMICS 37

20. A ship X moves due north at 30·0 km h^{-1}; a ship Y moves N. 60° W. at 20·0 km h^{-1}. Find the velocity of Y relative to X in magnitude and direction. If Y is 10 km due east of X at this instant, find the closest distance of approach of the two ships.

21. Two buckets of mass 6 kg are each attached to one end of a long inextensible string passing over a fixed pulley. If a 2 kg mass of putty is dropped from a height of 5 m into one bucket, calculate (i) the initial velocity of the system, (ii) the acceleration of the system, (iii) the loss of energy of the 2 kg mass due to the impact.

22. A bullet of mass 25 g and travelling horizontally at a speed of 200 m s^{-1} imbeds itself in a wooden block of mass 5 kg suspended by cords 3 m long. How far will the block swing from its position of rest before beginning to return? Describe a suitable method of suspending the block for this experiment and explain briefly the principles used in the solution of the problem. (*L.*)

23. State the principle of the conservation of linear momentum and show how it follows from Newton's laws of motion.
A stationary radioactive nucleus of mass 210 units disintegrates into an alpha particle of mass 4 units and a residual nucleus of mass 206 units. If the kinetic energy of the alpha particle is *E*, calculate the kinetic energy of the residual nucleus. (*N.*)

24. Define linear momentum and state the principle of conservation of linear momentum. Explain briefly how you would attempt to verify this principle by experiment.
Sand is deposited at a uniform rate of 20 kilogramme per second and with negligible kinetic energy on to an empty conveyor belt moving horizontally at a constant speed of 10 metre per minute. Find (*a*) the force required to maintain constant velocity, (*b*) the power required to maintain constant velocity, and (*c*) the rate of change of kinetic energy of the moving sand. Why are the latter two quantities unequal? (*O. & C.*)

25. What do you understand by the *conservation of energy*? Illustrate your answer by reference to the energy changes occurring (*a*) in a body whilst falling to and on reaching the ground, (*b*) in an X-ray tube.
The constant force resisting the motion of a car, of mass 1,500 kg, is equal to one-fifteenth of its weight. If, when travelling at 48 km per hour, the car is brought to rest in a distance of 50 m by applying the brakes, find the additional retarding force due to the brakes (assumed constant) and the heat developed in the brakes. (*N.*)

26. Define *uniform acceleration*. State, for each case, *one* set of conditions sufficient for a body to describe (*a*) a parabola, (*b*) a circle.
A projectile is fired from ground level, with velocity 500 m s^{-1} at 30° to the horizontal. Find its horizontal range, the greatest vertical height to which it rises, and the time to reach the greatest height. What is the least speed with which it could be projected in order to achieve the same horizontal range? (The resistance of the air to the motion of the projectile may be neglected.) (*O.*)

27. Define *momentum* and state the *law of conservation of linear momentum*. Discuss the conservation of linear momentum in the following cases (*a*) a freely falling body strikes the ground without rebounding, (b) during free flight an explosive charge separates an earth satellite from its propulsion unit, (c) a billiard ball bounces off the perfectly elastic cushion of a billiard table.

A bullet of mass 10 g travelling horizontally with a velocity of 300 m s^{-1} strikes a block of wood of mass 290 g which rests on a rough horizontal floor. After impact the block and bullet move together and come to rest when the block has travelled a distance of 15 m. Calculate the coefficient of sliding friction between the block and the floor. (*O. & C.*)

28. Explain the distinction between *fundamental* and *derived* units, using two examples of each.

Derive the dimensions of (a) *the moment of a couple* and *work*, and comment on the results, (b) the constants a and b in van der Waals' equation $(p + a/v^2)(v - b) = rT$ for unit mass of a gas. (*N.*)

29. Explain what is meant by the relative velocity of one moving object with respect to another.

A ship A is moving eastward with a speed of 15 km h^{-1} and another ship B, at a given instant 10 km east of A, is moving southwards with a speed of 20 km h^{-1}. How long after this instant will the ships be nearest to each other, how far apart will they be then, and in what direction will B be sighted from A? (*C.*)

30. Define *momentum* and state the *law of conservation of linear momentum*.

Outline an experiment to demonstrate momentum conservation and discuss the accuracy which could be achieved.

Show that in a collision between two moving bodies in which no external forces act, the conservation of linear momentum may be deduced directly from Newton's laws of motion.

A small spherical body slides with velocity v and without rolling on a smooth horizontal table and collides with an identical sphere which is initially at rest on the table. After the collision the two spheres slide without rolling away from the point of impact, the velocity of the first sphere being in a direction at 30° to its previous velocity. Assuming that energy is conserved, and that there are no horizontal external forces acting, calculate the speed and direction of travel of the target sphere away from the point of impact. (*O. & C.*)

31. Answer the following questions making particular reference to the physical principles concerned: (a) explain why the load on the back wheels of a motor car increases when the vehicle is accelerating, (b) the diagram, Fig. 1.23, shows a painter in a crate which hangs alongside a building. When the painter who weighs 1000 N pulls on the rope the force he exerts on the floor of the crate is 450 N. If the crate weighs 250 N find the acceleration. (*N.*)

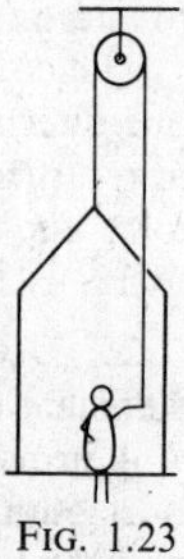

32. Derive an expression for the kinetic energy of a moving body.

FIG. 1.23

A vehicle of mass 2000 kg travelling at 10 m s^{-1} on a horizontal surface is brought to rest in a distance of 12·5 m by the action of its brakes. Calculate the average retarding force. What horse-power must the engine develop in order to take the vehicle up an incline of 1 in 10 at a constant speed of 10 m s^{-1} if the frictional resistance is equal to 200 N? (*L.*)

DYNAMICS

33. Explain what is meant by the principle of conservation of energy for a system of particles not acted upon by any external forces. What modifications are introduced when external forces are operative?

A bobsleigh is travelling at 10 m s^{-1} when it starts ascending an incline of 1 in 100. If it comes to rest after travelling 150 m up the slope, calculate the proportion of the energy lost in friction and deduce the coefficient of friction between the runners and the snow. (*O. & C.*)

34. State Newton's Laws of Motion and deduce from them the relation between the distance travelled and the time for the case of a body acted upon by a constant force. Explain the units in which the various quantities are measured.

A fire engine pumps water at such a rate that the velocity of the water leaving the nozzle is 15 m s^{-1}. If the jet be directed perpendicularly on to a wall and the rebound of the water be neglected, calculate the pressure on the wall (1 m^3 water weighs 1000 kg). (*O. & C.*)

Chapter 2

CIRCULAR MOTION. SIMPLE HARMONIC MOTION. WAVES. GRAVITATION

Angular Velocity

IN the previous chapter we discussed the motion of an object moving in a straight line. There are numerous cases of objects moving in a curve about some fixed point. The earth and the moon revolve continuously round the sun, for example, and the rim of the balance-wheel of a watch moves to-and-fro in a circular path about the fixed axis of the wheel. In this chapter we shall study the motion of an object moving in a circle with a *uniform speed* round a fixed point O as centre, Fig. 2.1.

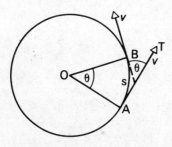

FIG. 2.1 Circular motion

If the object moves from A to B so that the radius OA moves through an angle θ, its *angular velocity*, ω, about O is defined as the *change of the angle per second*. Thus if t is the time taken by the object to move from A to B,

$$\omega = \frac{\theta}{t} . \quad . \quad . \quad . \quad (1)$$

Angular velocity is usually expressed in 'radian per second' (rad s^{-1}). From (1),

$$\theta = \omega t . \quad . \quad . \quad . \quad (2)$$

which is analogous to the formula 'distance = uniform velocity × time' for motion in a straight line. It will be noted that the time T to describe the circle once, known as the *period* of the motion, is given by

$$T = \frac{2\pi}{\omega}, \quad . \quad . \quad . \quad (3)$$

since 2π radians = 360° by definition.

If s is the length of the arc AB, then $s/r = \theta$, by definition of an angle in radians.

$$\therefore s = r\theta.$$

Dividing by t, the time taken to move from A to B,

$$\therefore \frac{s}{t} = r\frac{\theta}{t}.$$

But s/t = the *speed*, v, of the rotating object, and θ/t is the angular velocity.

$$\therefore v = r\omega \qquad . \qquad . \qquad . \qquad . \qquad (4)$$

Acceleration in a circle

When a stone is attached to a string and whirled round at constant speed in a circle, one can feel the force in the string needed to keep the stone moving. The presence of the force, called a *centripetal force*, implies that the stone has an acceleration. And since the force acts towards the centre of the circle, the direction of the acceleration, which is a vector quantity, is also towards the centre.

To obtain an expression for the acceleration towards the centre, consider an object moving with a constant speed v round a circle of radius r. Fig. 2.2 (i). At A, its *velocity* v_A is in the direction of the tangent AC; a short time δt later at B, its velocity v_B is in the direction of the tangent BD. Since their directions are different, the velocity v_B is different from the velocity v_A, although their magnitudes are both equal to v. Thus a velocity change or acceleration has occurred from A to B.

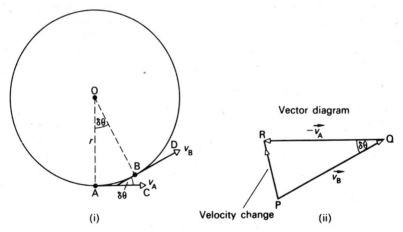

FIG. 2.2 Acceleration in circle

The velocity change from A to B = $\vec{v}_B - \vec{v}_A = \vec{v}_B + (-\vec{v}_A)$. The arrows denote vector quantities. In Fig. 2.2 (ii), PQ is drawn to represent

v_B in magnitude (v) and direction (BD); QR is drawn to represent $(-\vec{v}_A)$ in magnitude (v) and direction (CA). Then, as shown on p. 11,

$$\text{velocity change} = \vec{v}_B + (-\vec{v}_A) = \vec{PR}.$$

When δt is small, the angle AOB or $\delta \theta$ is small. Thus angle PQR, equal to $\delta \theta$, is small. PR then points towards O, the centre of the circle. *The velocity change or acceleration is thus directed towards the centre.*

The magnitude of the acceleration, a, is given by

$$a = \frac{\text{velocity change}}{\text{time}} = \frac{PR}{\delta t}.$$

$$= \frac{v \cdot \delta \theta}{\delta t}.$$

since $PR = v \cdot \delta \theta$. In the limit, when δt approaches zero, $\delta \theta / \delta t = d\theta/dt = \omega$, the angular velocity. But $v = r\omega$ (p. 41). Hence, since $a = v\omega$,

$$a = \frac{v^2}{r} \quad \text{or} \quad r\omega^2. \qquad . \qquad . \qquad (5)$$

Thus an object moving in a circle of radius r with a constant speed v has a constant acceleration towards the centre equal to v^2/r or $r\omega^2$.

Centripetal forces

The force F required to keep an object of mass m moving in a circle of radius $r = ma = mv^2/r$. It is called a *centripetal force* and acts *towards the centre* of the circle. When a stone A is whirled in a horizontal circle of centre O by means of a string, the tension T provides the centripetal force. Fig. 2.3 (i). For a racing car moving round a circular track, the friction at the wheels provides the centripetal force. Planets such as P, moving in a circular orbit round the sun S, have a centripetal force due to gravitational attraction between S and P (p. 66). Fig. 2.3 (ii).

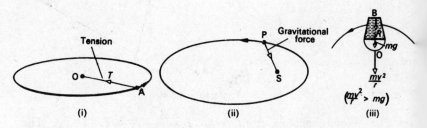

Fig. 2.3 Centripetal forces

If some water is placed in a bucket B attached to the end of a string, the bucket can be whirled in a vertical plane without any water falling out. When the bucket is vertically above the point of support O, the weight mg of the water is less than the required force mv^2/r towards the centre and so the water stays in. Fig. 2.3 (iii). The reaction R of the bucket base on the water provides the rest of the force. If the bucket is whirled slowly and $mg > mv^2/r$, part of the weight provides the force mv^2/r. The rest of the weight causes the water to accelerate downward and hence to leave the bucket.

Centrifuges

Centrifuges are used to separate particles in suspension from the more dense liquid in which they are contained. This mixture is poured into a tube in the centrifuge, which is then whirled at high speed in a horizontal circle.

The pressure gradient due to the surrounding liquid at a particular distance, r say, from the centre provides a centripetal force of $mr\omega^2$ for a small volume of liquid of mass m, where ω is the angular velocity. If the volume of liquid is replaced by an equal volume of particles of smaller mass m' than the liquid, the centripetal force acting on the particles at the same place is then greater than that required by $(m-m')r\omega^2$. The net force urges the particles towards the centre in spiral paths, and here they collect. Thus when the centrifuge is stopped, and the container or tube assumes a vertical position, the suspension is found at the top of the tube and clear liquid at the bottom. Conversely, particles suspended in a *less* dense liquid collect at the bottom of the tube in a centrifuge, leaving clear liquid at the top.

Motion of Bicycle Rider Round Circular Track

When a person on a bicycle rides round a circular racing track, the frictional force F at the ground provides the inward force towards the centre or centripetal force. Fig. 2.4. This produces a moment about his centre of gravity G which is counterbalanced, when he leans inwards, by the moment of the normal reaction R. Thus provided no skidding occurs, $F \cdot h = R \cdot a = mg \cdot a$, since $R = mg$ for no vertical motion.

$$\therefore \frac{a}{h} = \tan \theta = \frac{F}{mg},$$

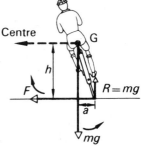

Fig. 2.4 Rider on circular track

where θ is the angle of inclination to the vertical. Now $F = mv^2/r$.

$$\therefore \tan\theta = \frac{v^2}{rg}.$$

When F is greater than the limiting friction, skidding occurs. In this case $F > \mu mg$, or $mg \tan\theta > \mu mg$. Thus $\tan\theta > \mu$ is the condition for skidding.

Motion of Car (or Train) Round Circular Track

Suppose a car (or train) is moving with a velocity v round a horizontal circular track of radius r, and let R_1, R_2 be the respective normal reactions at the wheels A, B, and F_1, F_2 the corresponding frictional forces, Fig. 2.5. Then, for circular motion we have

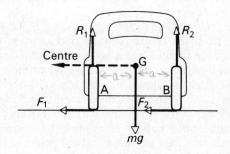

FIG. 2.5 Car on circular track

$$F_1 + F_2 = \frac{mv^2}{r}, \quad \ldots \quad \text{(i)}$$

and vertically $\quad R_1 + R_2 = mg. \quad \ldots \quad \text{(ii)}$

Also, taking moments about G,

$$(F_1 + F_2)h + R_1 a - R_2 a = 0 \quad \ldots \quad \text{(iii)}$$

where $2a$ is the distance between the wheels, assuming G is mid-way between the wheels, and h is the height of G above the ground. From these three equations, we find

$$R_2 = \tfrac{1}{2}m\left(g + \frac{v^2 h}{ra}\right)$$

and, vertically, $\quad R_1 = \tfrac{1}{2}m\left(g - \frac{v^2 h}{ra}\right).$

R_2 never vanishes since it always has a positive value. But if

$v^2 = arg/h$, $R_1 = 0$, and the car is about to overturn outwards. R_1 will be positive if $v^2 < arg/h$.

Motion of Car (or Train) Round Banked Track

Suppose a car (or train) is moving round a banked track in a circular path of horizontal radius r, Fig. 2.6. If the only forces at the wheels

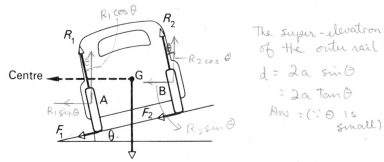

FIG. 2.6 Car on banked track

A, B are the normal reactions R_1, R_2 respectively, that is, there is no side-slip or strain at the wheels, the force towards the centre of the track is $(R_1 + R_2) \sin \theta$, where θ is the angle of inclination of the plane to the horizontal.

$$\therefore (R_1 + R_2) \sin \theta = \frac{mv^2}{r} \qquad \text{(i)}$$

For vertical equilibrium, $(R_1 + R_2) \cos \theta = mg$ . . . (ii)

Dividing (i) by (ii), $\therefore \tan \theta = \dfrac{v^2}{rg}$ . . . (iii)

Thus for a given velocity v and radius r, the angle of inclination of the track for no side-slip must be $\tan^{-1}(v^2/rg)$. As the speed v increases, the angle θ increases, from (iii). A racing-track is made saucer-shaped because at higher speeds the cars can move towards a part of the track which is steeper and sufficient to prevent side-slip. The outer rail of a curved railway track is raised about the inner rail so that the force towards the centre is largely provided by the component of the reaction at the wheels. It is desirable to bank a road at corners for the same reason as a racing track is banked.

Thrust at Ground

Suppose now that the car (or train) is moving at such a speed that the frictional forces at A, B are F_1, F_2 respectively, each acting towards the centre of the track.

Resolving horizontally,

$$\therefore (R_1+R_2)\sin\theta + (F_1+F_2)\cos\theta = \frac{mv^2}{r} \qquad (i)$$

Resolving vertically,

$$\therefore (R_1+R_2)\cos\theta - (F_1+F_2)\sin\theta = mg \qquad (ii)$$

Solving, we find

$$F_1+F_2 = m\left(\frac{v^2}{r}\cos\theta - g\sin\theta\right) \qquad (iii)$$

If $\frac{v^2}{r}\cos\theta > g\sin\theta$, then (F_1+F_2) is positive; and in this case both the thrusts on the wheels at the ground are towards the centre of the track.

If $\frac{v^2}{r}\cos\theta < g\sin\theta$, then (F_1+F_2) is negative. In this case the forces F_1 and F_2 act outwards away from the centre of the track.

For stability, we have, by moments about G,

$$(F_1+F_2)h + R_1 a - R_2 a = 0$$

$$\therefore (F_1+F_2)\frac{h}{a} = R_2 - R_1.$$

From (iii),
$$\therefore \frac{mh}{a}\left(\frac{v^2}{r}\cos\theta - g\sin\theta\right) = R_2 - R_1 \qquad (iv)$$

The reactions R_1, R_2 can be calculated by finding (R_1+R_2) from equations (i), (ii), and combining the result with equation (iv). This is left as an exercise to the student.

Variation of g with latitude

The acceleration due to gravity, g, varies over the earth's surface. This is due to two main causes. Firstly, the earth is elliptical, with the polar radius, b, $6·357 \times 10^6$ metre and the equatorial radius, a, $6·378 \times 10^6$ metre, and hence g is greater at the poles than at the equator, where the body is further away from the centre of the earth. Secondly, the earth rotates about the polar axis, AB. Fig. 2.7. We shall consider this effect in more detail, and suppose the earth is a perfect sphere.

In general, an object of mass m suspended by a spring-balance at a point on the earth would be acted on by an upward force $T = mg'$, where g' is the observed or apparent acceleration due to gravity. There would also be a downward attractive force mg towards the centre of the earth, where g is the acceleration in the absence of rotation.

(1) *At the poles*, A or B, there is no rotation. Hence $mg - T = 0$, or $mg = T = mg'$. Thus $g' = g$.

(2) *At the equator*, C or D, there is a resultant force $mr\omega^2$ towards the centre where r is the earth's radius. Since OD is the vertical, we have

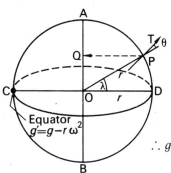

FIG. 2.7 Variation of g

$$mg - T = mr\omega^2.$$
$$\therefore T = mg - mr\omega^2 = mg'$$
$$\therefore g' = g - r\omega^2.$$

The radius r of the earth is about 6.37×10^6 m, and $\omega = [(2\pi/(24 \times 3600)]$ radian per second.

$$\therefore g - g' = r\omega^2 = \frac{6.37 \times 10^6 \times (2\pi)^2}{(24 \times 3600)^2} = 0.034.$$

Latest figures give g, at the pole, 9·832 m s^{-2}, and g', at the equator, 9·780 m s^{-2}, a difference of 0·052 m s^{-2}. The earth's rotation accounts for 0·034 m s^{-2}.

(3) *At latitude* λ. Consider an object suspended by a string at P on the earth, where the latitude is λ. Fig. 2.7. The resultant force is directed along PQ, and is equal to $m.r\cos\lambda.\omega^2$, since PQ $= r\cos\lambda$. The string suspending the object is now inclined at a very small angle θ to the vertical OP. For motion in a circle of radius PQ, we have

$$mg\cos\lambda - T\cos(\lambda+\theta) = mr\omega^2\cos\lambda \quad . \quad . \quad . \quad (i)$$

and for no motion in a perpendicular direction,

$$mg\sin\lambda - T\sin(\lambda+\theta) = 0 \quad . \quad . \quad . \quad (ii)$$

From (i), $\qquad T\cos(\lambda+\theta) = mg\cos\lambda - mr\omega^2\cos\lambda.$

From (ii), $\qquad T\sin(\lambda+\theta) = mg\sin\lambda.$

Squaring and adding,

$$\therefore T^2 = (mg\cos\lambda - mr\omega^2\cos\lambda)^2 + (mg\sin\lambda)^2 = (mg')^2.$$
$$\therefore g'^2 = g^2 - 2gr\omega^2\cos^2\lambda + r^2\omega^4\cos^2\lambda.$$
$$\therefore g' = [g^2 - 2gr\omega^2\cos^2\lambda + r^2\omega^4\cos^2\lambda]^{1/2}.$$
$$= g\left[1 - \frac{2r\omega^2\cos^2\lambda}{g} + \frac{r^2\omega^4\cos^2\lambda}{g^2}\right]^{1/2}.$$

Neglecting $r^2\omega^4\cos^2\lambda/g^2$, which is very small, and expanding by the binomial theorem, we have

$$g' = g\left(1 - \frac{r\omega^2\cos^2\lambda}{g}\right) = g - r\omega^2\cos^2\lambda.$$

$\therefore g - g' =$ reduction in acceleration of gravity $= r\omega^2\cos^2\lambda.$

An accurate formula for the variation of g' with latitude, recommended by an international meeting at Stockholm in 1930, is

$$g' = 980·6294 - 2·5862 \cos 2\lambda + 0·0058 \cos^2 2\lambda (\times 10^{-2} \text{ m s}^{-2}).$$

Substituting for T from (ii) in (i) on p. 47, then

$$mg \cos \lambda - \frac{mg \sin \lambda \cos (\lambda + \theta)}{\sin (\lambda + \theta)} = mr\omega^2 \cos \lambda.$$

$$\therefore \frac{g \sin \theta}{\sin (\lambda + \theta)} = r\omega^2 \cos \lambda.$$

As θ is very small compared with λ, $\sin(\lambda + \theta) = \sin \lambda$ to a good approximation.

$$\therefore \sin \theta = \frac{r\omega^2 \cos \lambda . \sin \lambda}{g} = \frac{r\omega^2}{2g} \sin 2\lambda.$$

Thus θ is greatest when $\lambda = 45°$; in this case θ is about $\frac{1}{10}°$. This is the angle which a plumbline would make with the vertical owing to the earth's rotation.

EXAMPLE

Explain the action of a centrifuge when used to hasten the deposition of a sediment from a liquid.

A pendulum bob of mass 1 kg is attached to a string 1 m long and made to revolve in a horizontal circle of radius 60 cm. Find the period of the motion and the tension of the string. (C.)

First part. See text, p. 43.

Second part. Suppose A is the bob, and OA is the string, Fig. 2.8. If T is the tension in newtons, and θ is the angle of inclination of OA to the horizontal, then, for motion in the circle of radius $r = 60$ cm $= 0·6$ m,

$$T \cos \theta = \frac{mv^2}{r} = \frac{mv^2}{0·6} \qquad . \qquad . \qquad \text{(i)}$$

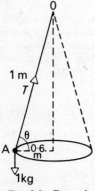

FIG. 2.8 Example

Since the bob A does not move in a vertical direction, then

$$T \sin \theta = mg \qquad . \qquad . \qquad . \qquad . \qquad \text{(ii)}$$

Now $\cos \theta = \frac{60}{100} = \frac{3}{5}$; hence $\sin \theta = \frac{4}{5}$.

From (ii),

$$\therefore T = \frac{mg}{\sin \theta} = \frac{1 \times 9·8}{4/5} = 12·25 \text{ N}.$$

From (i)

$$v = \sqrt{\frac{0·6T \cos \theta}{m}}$$

$$= \sqrt{\frac{0·6 \times 12·25 \times 3}{1 \times 5}} = 2·1 \text{ m s}^{-1}$$

∴ angular velocity, $\omega = \dfrac{v}{r} = \dfrac{2 \cdot 1}{0 \cdot 6} = \dfrac{7}{2}$ rad s^{-1}

∴ period, $T, = \dfrac{2\pi}{\omega} = \dfrac{2\pi}{7/2} = \dfrac{4\pi}{7}$ second.

∴ $T = 1 \cdot 8$ second.

SIMPLE HARMONIC MOTION

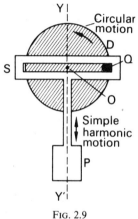

FIG. 2.9
Simple harmonic motion

When the bob of a pendulum moves to-and-fro through a small angle, the bob is said to be moving with *simple harmonic motion*. The prongs of a sounding tuning fork, and the layers of air near it, are moving with simple harmonic motion, and light waves can be considered due to simple harmonic variations.

Simple harmonic motion is closely associated with circular motion. An example is shown in Fig. 2.9. This illustrates an arrangement used to convert the circular motion of a disc D into the to-and-fro or simple harmonic motion of a piston P. The disc is driven about its axle O by a peg Q fixed near its rim. The vertical motion drives P up and down. Any horizontal component of the motion merely causes Q to move along the slot S. Thus the simple harmonic motion of P is the *projection* on the vertical line YY′ of the circular motion of Q.

An everyday example of an opposite conversion of motion occurs in car engines. Here the to-and-fro or 'reciprocating' motion of the piston engine is changed to a regular circular motion by connecting rods and shafts so that the wheels are turned.

Formulae in Simple Harmonic Motion

Consider an object moving round a circle of radius r and centre Z with a uniform angular velocity ω, Fig. 2.10. If CZF is a fixed diameter, the *foot* of the perpendicular from the moving object to this diameter moves from Z to C, back to Z and across to F, and then returns to Z, while the object moves once round the circle from O in an anti-clockwise direction. The to-and-fro motion along CZF of the foot of the perpendicular is defined as *simple harmonic motion*.

Suppose the object moving round the circle is at A at some instant, where angle OZA = θ, and suppose the foot of the perpendicular from A to CZ is M. The acceleration of the object at A is $\omega^2 r$, and this

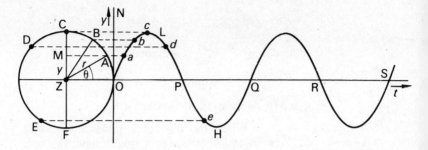

FIG. 2.10 Simple harmonic curve

acceleration is directed along the radius AZ (see p. 42). Hence the acceleration of M towards Z

$$= \omega^2 r \cos AZC = \omega^2 r \sin \theta.$$

But $r \sin \theta = MZ = y$ say.

∴ acceleration of M towards Z $= \omega^2 y$.

Now ω^2 is a constant.

∴ *acceleration of M towards Z $\propto$ distance of M from Z.*

If we wish to express mathematically that the acceleration is always directed towards Z, we must say

$$\text{acceleration towards Z} = -\omega^2 y \qquad . \qquad . \quad (1)$$

The minus indicates, of course, that the object begins to retard as it passes the centre, Z, of its motion. If the minus were omitted from equation (1) the latter would imply that the acceleration increases as y increases, and the object would then never return to its original position.

We can now form a definition of simple harmonic motion. It is the motion of a particle *whose acceleration is always (i) directed towards a fixed point, (ii) directly proportional to its distance from that point.*

Period, Amplitude. Sine Curve

The time taken for the foot of the perpendicular to move from C to F and back to C is known as the *period* (T) of the simple harmonic motion. In this time, the object moving round the circle goes exactly once round the circle from C; and since ω is the angular velocity and 2π radians (360°) is the angle described, the period T is given by

$$T = \frac{2\pi}{\omega} \qquad . \qquad . \qquad . \quad (1)$$

SIMPLE HARMONIC MOTION

The distance ZC, or ZF, is the maximum distance from Z of the foot of the perpendicular, and is known as the *amplitude* of the motion. It is equal to r, the radius of the circle.

We have now to consider the variation with time, t, of the distance, y, from Z of the foot of the perpendicular. The distance $y = ZM = r \sin \theta$. But $\theta = \omega t$, where ω is the angular velocity.

$$\therefore y = r \sin \omega t \quad . \quad . \quad . \quad . \quad (2)$$

The graph of y v. t is shown in Fig. 2.10, where ON represents the y-axis and OS the t-axis; since the angular velocity of the object moving round the circle is constant, θ is proportional to the time t. Thus as the foot of the perpendicular along CZF moves from Z to C and back to Z, the graph OLP is traced out; as the foot moves from Z to F and returns to Z, the graph PHQ is traced out. The graph is a *sine curve*. The complete set of values of y from O to Q is known as a cycle. The number of cycles per second is called the *frequency*. The unit '1 cycle per second' is called '1 *hertz (Hz)*'. The mains frequency in Great Britain is 50 Hz or 50 cycles per second.

Velocity during S.H.M.

Suppose the object moving round the circle is at A at some instant, Fig. 2.10. The velocity of the object is $r\omega$, where r is the radius of the circle, and it is directed along the tangent at A. Consequently the velocity parallel to the diameter FC at this instant = $r\omega \cos \theta$, by resolving.

$$\therefore \text{velocity, } v, \text{ of M along FC} = r\omega \cos \theta.$$

But $\qquad y = r \sin \theta$

$$\therefore \cos \theta = \sqrt{1 - \sin^2 \theta} = \sqrt{1 - y^2/r^2} = \frac{1}{r}\sqrt{r^2 - y^2}$$

$$\therefore v = \omega \sqrt{r^2 - y^2} \quad . \quad . \quad . \quad . \quad . \quad (1)$$

This is the expression for the velocity of an object moving with simple harmonic motion. The maximum velocity, v_m, corresponds to $y = 0$, and hence

$$v_m = \omega r. \quad . \quad . \quad . \quad (2)$$

Summarising our results:

(1) If the acceleration a of an object = $-\omega^2 y$, where y is the distance or displacement of the object from a fixed point, the motion is simple harmonic motion.

(2) The *period*, T, of the motion $= 2\pi/\omega$, where T is the time to make a complete to-and-fro movement or cycle. The *frequency*, f, $= 1/T$ and its unit is 'Hz'.

(3) The amplitude, r, of the motion is the maximum distance on either side of the centre of oscillation.

(4) The velocity at any instant, $v, = \omega\sqrt{r^2 - y^2}$; the maximum velocity $= \omega r$. Fig. 2.11 (i) shows a graph of the variation of v and acceleration a with displacement y, which are respectively an ellipse and a straight line.

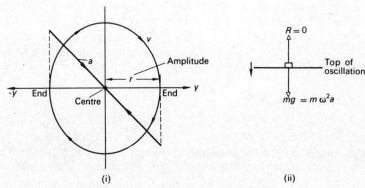

FIG. 2.11 Simple harmonic motion

S.H.M. and g

If a small coin is placed on a horizontal platform connected to a vibrator, and the amplitude is kept constant as the frequency is increased from zero, the coin will be heard 'chattering' at a particular frequency f_0. At this stage the reaction of the table with the coin becomes zero at some part of every cycle, so that it loses contact periodically with the surface. Fig. 2.11 (ii).

The maximum acceleration in S.H.M. occurs at the end of the oscillation because the acceleration is directly proportional to the displacement. Thus maximum acceleration $= \omega^2 a$, where a is the amplitude and ω is $2\pi f_0$.

The coin will lose contact with the table when it is moving *down* with acceleration g (Fig. 2.11 (ii)). Suppose the amplitude is 8·0 cm. Then

$$(2\pi f_0)^2 a = g$$
$$\therefore 4\pi^2 f_0^2 \times 0{\cdot}08 = 9{\cdot}8$$

$$\therefore f_0 = \sqrt{\frac{9 \cdot 8}{4\pi^2 \times 0 \cdot 08}} = 1 \cdot 8 \text{ Hz.}$$

Damping of S.H.M.

In practice, simple harmonic variations of a pendulum, for example, will die away as the energy is dissipated by viscous forces due to the air. The oscillation is then said to be *damped*. In the absence of any damping forces the oscillations are said to be *free*.

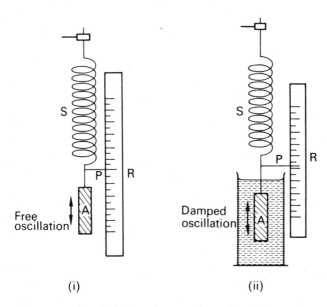

FIG. 2.12 Experiment on damped oscillations

A simple experiment to investigate the effect of damping is illustrated in Fig. 2.12 (i). A suitable weight A is suspended from a helical spring S, a pointer P is attached to S, and a vertical scale R is set up behind P. The weight A is then set pulled down and released. The period, and the time taken for the oscillations to die away, are noted.

As shown in Fig. 2.12 (ii), A is now fully immersed in a damping medium, such as a light oil, water or glycerine. A is then set oscillating, and the time for oscillations to die away is noted. It is shorter than before and least for the case of glycerine. The decreasing amplitude in successive oscillations may also be noted from the upward limit of travel of P and the results plotted.

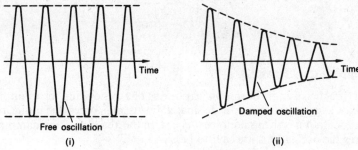

FIG. 2.13 Free and damped oscillations

Fig. 2.13 (i), (ii) shows how damping produces an exponential fall in the amplitude with time.

The experiment works best for a period of about $\frac{1}{2}$-second and a weight which is long and thin so that the damping is produced by non-turbulent fluid flow over the vertical sides. During the whole cycle, A must be totally immersed in the fluid.

EXAMPLE

A steel strip, clamped at one end, vibrates with a frequency of 20 Hz and an amplitude of 5 mm at the free end, where a small mass of 2 g is positioned. Find (a) the velocity of the end when passing through the zero position, (b) the acceleration at maximum displacement, (c) the maximum kinetic and potential energy of the mass.

Suppose $y = r \sin \omega t$ represents the vibration of the strip where r is the amplitude.

(a) The velocity, $v, = \omega \sqrt{r^2 - y^2}$ (p. 52). When the end of the strip passes through the zero position $y = 0$; and the maximum speed, v_m, is given by

$$v_m = \omega r.$$

Now $\omega = 2\pi f = 2\pi \times 20$, and $r = 0.005$ m.

$$\therefore v_m = 2\pi \times 20 \times 0.005 = 0.628 \text{ m s}^{-1}.$$

(b) The acceleration $= -\omega^2 y = -\omega^2 r$ at the maximum displacement.

$$\therefore \text{acceleration} = (2\pi \times 20)^2 \times 0.005$$
$$= 79 \text{ m s}^{-2}.$$

(c) $m = 2$ g $= 2 \times 10^{-3}$ kg, $v_m = 0.628$ m s^{-1}.

$$\therefore \text{maximum K.E.} = \tfrac{1}{2} m v_m^2 = \tfrac{1}{2} \times (2 \times 10^{-3}) \times 0.628^2 = 3.9 \times 10^{-4} \text{ J (approx.).}$$

Maximum P.E. ($v = 0$) = Maximum K.E. = 3.9×10^{-4} J.

Simple Pendulum

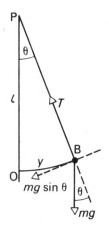

FIG. 2.14
Simple pendulum

We shall now study some cases of simple harmonic motion. Consider a *simple pendulum*, which consists of a small mass *m* attached to the end of a length *l* of wire, Fig. 2.14. If the other end of the wire is attached to a fixed point P and the mass is displaced slightly, it oscillates to-and-fro along the arc of a circle of centre P. We shall now show that the motion of the mass about its original position O is simple harmonic motion.

Suppose that the vibrating mass is at B at some instant, where OB = y and angle OPB = θ. At B, the force pulling the mass towards O is directed along the tangent at B, and is equal to $mg \sin \theta$. The tension, T, in the wire has no component in this direction, since PB is perpendicular to the tangent at B. Thus, since force = mass × acceleration (p. 14),

$$-mg \sin \theta = ma,$$

where a is the acceleration along the arc OB; the minus indicates that the force is towards O, while the displacement, y, is measured along the arc from O in the opposite direction. *When θ is small, $\sin \theta = \theta$ in radians*; also $\theta = y/l$. Hence,

$$-mg\theta = -mg\frac{y}{l} = ma$$

$$\therefore a = -\frac{g}{l}y = -\omega^2 y,$$

where $\omega^2 = g/l$. Since the acceleration is proportional to the distance y from a fixed point, the motion of the vibrating mass is simple harmonic motion (p. 50). Further, from p. 50, the period $T = 2\pi/\omega$.

$$\therefore T = \frac{2\pi}{\sqrt{g/l}} = 2\pi\sqrt{\frac{l}{g}} \qquad . \qquad . \qquad . \qquad (1)$$

At a given place on the earth, where g is constant, the formula shows that the period T depends only on the length, l, of the pendulum. Moreover, the period remains constant even when the amplitude of the vibrations diminish owing to the resistance of the air. This result was first obtained by Galileo, who noticed a swinging lantern one day, and timed the oscillations by his pulse (there were no clocks in his day). He found that the period remained constant although the swings gradually diminished in amplitude.

Determination of g by Simple Pendulum

The acceleration due to gravity, g, can be found by measuring the period, T, of a simple pendulum corresponding to a few different lengths, l, from 80 cm to 180 cm for example. To perform the experiment accurately: (i) Fifty oscillations should be timed, (ii) a small angle of swing is essential, less than 10°, (iii) a small sphere should be tied to the end of a thread to act as the mass, and its radius added to the length of the thread to determine l.

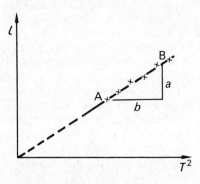

FIG. 2.15   Graph of l v. T^2

A graph of l against T^2 is now plotted from the results, and a straight line AB, which should pass through the origin, is then drawn to lie evenly between the points, Fig. 2.15.

Now
$$T = 2\pi\sqrt{\frac{l}{g}},$$

$$\therefore T^2 = \frac{4\pi^2 l}{g}$$

$$\therefore g = 4\pi^2 \times \frac{l}{T^2}. \quad . \quad . \quad . \quad (1)$$

The gradient a/b of the line AB is the magnitude of l/T^2; and by substituting in (1), g can then be calculated.

If the pendulum is suspended from the ceiling of a very tall room and the string and bob reaches nearly to the floor, then one may proceed to find g by (i) measuring the period T_1, (ii) cutting off a measured length a of the string and determining the new period T_2 with the shortened string. Then, if h is the height of the ceiling above the bob initially, $T_1 = 2\pi\sqrt{h/g}$ and $T_2 = 2\pi\sqrt{(h-a)/g}$. Thus

$$h = \frac{gT_1^2}{4\pi^2} \quad \text{and} \quad h-a = \frac{gT_2^2}{4\pi^2}.$$

$$\therefore a = \frac{g}{4\pi^2}(T_1^2 - T_2^2).$$

$$\therefore g = \frac{4\pi^2 a}{T_1^2 - T_2^2}.$$

SIMPLE HARMONIC MOTION

Thus g can be calculated from a, T_1 and T_2. Alternatively, the period T can be measured for several lengths a. Then, since $T = 2\pi\sqrt{(h-a)/g}$,

$$h - a = \frac{g}{4\pi^2}T^2.$$

A graph of a v. T^2 is thus a straight line whose gradient is $g/4\pi^2$. Hence g can be found. The intercept on the axis of a, when $T^2 = 0$, is h, the height of the ceiling above the bob initially.

The Spiral Spring or Elastic Thread

When a weight is suspended from the end of a spring or an elastic thread, experiment shows that the extension of the spring, i.e., the increase in length, is proportional to the weight, provided that the elastic limit of the spring is not exceeded (see p. 181). Generally, then, *the tension (force), T, in a spring is proportional to the extension x produced*, i.e., $T = kx$, where k is a constant of the spring.

Consider a spring or an elastic thread PA of length l suspended from a fixed point P, Fig. 2.16. When a mass m is placed on it, the spring stretches to O by a length e given by

$$mg = ke, \quad . \quad . \quad . \quad \text{(i)}$$

since the tension in the spring is then mg. If the mass is pulled down a little and then released, it vibrates up-and-down above and below O. Suppose at an instant that B is at a distance x below O. The tension T of the spring at B is then equal to $k(e+x)$, and hence the force towards O $= k(e+x) - mg$. Since force $=$ mass $\times$ acceleration,

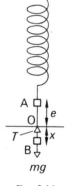

FIG. 2.16
Spiral spring

$$\therefore -[k(e+x) - mg] = ma,$$

the minus indicates that the net force is upward at this instant, whereas the displacement x is measured from O in the opposite direction at the same instant. From this equation,

$$-ke - kx + mg = ma.$$

But, from (i), $$mg = ke,$$

$$\therefore -kx = ma,$$

$$\therefore a = -\frac{k}{m}x = -\omega^2 x,$$

where $\omega^2 = k/m$. Thus the motion is simple harmonic about O, and the period T is given by

$$T = \frac{2\pi}{\omega} = 2\pi\sqrt{\frac{m}{k}} \ . \qquad . \qquad . \qquad . \qquad (1)$$

Also, since $mg = ke$, it follows that $m/k = e/g$.

$$\therefore T = 2\pi\sqrt{\frac{e}{g}} \qquad . \qquad . \qquad . \qquad . \qquad (2)$$

From (1), it follows that $T^2 = 4\pi^2 m/k$. Consequently a graph of T^2 v. m should be a straight line passing through the origin. In practice, when the load m is varied and the corresponding period T is measured, a straight line graph is obtained when T^2 is plotted against m, thus verifying indirectly that the motion of the load was simple harmonic. The graph does not pass through the origin, however, owing to the mass and the movement of the various parts of the spring. This has not been taken into account in the foregoing theory and we shall now show how g may be found in this case.

Determination of g by Spiral Spring

The mass s of a vibrating spring is taken into account, in addition to the mass m suspended at the end, theory beyond the scope of this book then shows that the period of vibration, T, is given by

$$T = 2\pi\sqrt{\frac{m + \lambda s}{k}} \qquad . \qquad . \qquad . \qquad (i)$$

where λ is approximately $\frac{1}{3}$ and k is the elastic constant of the spring. Squaring (i) and re-arranging,

$$\therefore \frac{k}{4\pi^2}T^2 = m + \lambda s \qquad . \qquad . \qquad . \qquad (ii)$$

Thus, since λ, k, s are constants, a graph of T^2 v. m should be a straight line when m is varied and T observed. A straight line graph verifies indirectly that the motion of the mass at the end of the spring is simple harmonic. Further, the magnitude of $k/4\pi^2$ can be found from the slope of the line, and hence k can be calculated.

If a mass M is placed on the end of the spring, producing a steady extension e less than the elastic limit, then $Mg = ke$.

$$\therefore g = \frac{e}{M} \times k \ . \qquad . \qquad . \qquad . \qquad (iii)$$

By attaching different masses to the spring, and measuring the corresponding extension, the magnitude of e/M can be found by plotting e v. M and measuring the slope of the line. This is called the 'static' experiment on the spring. From the magnitude of k obtained in the 'dynamic' experiment when the period was determined for different loads, the value of g can be found by substituting the magnitudes of e/M and k in (iii).

Oscillations of a Liquid in a U-Tube

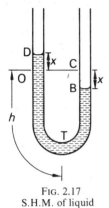

FIG. 2.17
S.H.M. of liquid

If the liquid on one side of a U-tube T is depressed by blowing gently down that side, the levels of the liquid will oscillate for a short time about their respective initial positions O, C, before finally coming to rest, Fig. 2.17.

The period of oscillation can be found by supposing that the level of the liquid on the left side of T is at D at some instant, at a height x above its original (undisturbed) position O. The level B of the liquid on the other side is then at a depth x below its original position C, and hence the excess pressure on the whole liquid, as shown on p. 125,

$$= \text{excess height} \times \text{density of liquid} \times g$$

$$= 2x\rho g.$$

Now pressure = force per unit area.
∴ force on liquid = pressure × area of cross-section of the tube

$$= 2x\rho g \times A,$$

where A is the cross-sectional area of the tube.

This force causes the liquid to accelerate. The mass of liquid in the U-tube = volume × density = $2hA\rho$, where $2h$ is the total length of the liquid in T. Now the acceleration, a, towards O or C is given by *force = mass × a*.

$$\therefore -2x\rho g A = 2hA\rho a.$$

The minus indicates that the force towards O is opposite to the displacement measured from O at that instant.

$$\therefore a = -\frac{g}{h}x = -\omega^2 x,$$

where $\omega^2 = \frac{g}{h}$. The motion of the liquid about O (or C) is thus simple harmonic, and the period T is given by

$$T = \frac{2\pi}{\omega} = 2\pi\sqrt{\frac{h}{g}}.$$

P.E. and K.E. exchanges in oscillating systems

We can now make a general point about *oscillations* and *oscillating systems*. As an illustration, suppose that one end of a spring S of negligible mass is attached to a smooth object A, and that S and A are laid on a horizontal smooth table. If the free end of S is attached to the table and A is pulled slightly to extend the spring and then released, the system vibrates with simple harmonic motion. This is the case discussed on p. 57, without taking gravity into account. The centre of oscillation O is the position of the end of the spring corresponding to its natural length, that is, when the spring is neither extended nor compressed. If the spring extension obeys the law *force* = kx, where k is a constant, and m is the mass of A, then, as on p. 58, it can easily be shown that the period T of oscillation is given by:

$$T = \frac{2\pi}{\omega} = 2\pi \sqrt{\frac{m}{k}}.$$

The energy of the stretched spring is *potential energy*, P.E.—its molecules are continually displaced or compressed relative to their normal distance apart. The P.E. for an extension $x = \int F \cdot dx = \int kx \cdot dx = \frac{1}{2}kx^2$.

The energy of the mass is *kinetic energy*, K.E., or $\frac{1}{2}mv^2$, where v is the velocity. Now from $x = a \sin \omega t$, $v = dx/dt = \omega a \cos \omega t$.

$\therefore$ total energy of spring plus mass = $\frac{1}{2}kx^2 + \frac{1}{2}mv^2$

$\qquad = \frac{1}{2}ka^2 \sin^2 \omega t + \frac{1}{2}m\omega^2 a^2 \cos^2 \omega t.$

But $\omega^2 = k/m$, or $k = m\omega^2$.

$\therefore$ total energy = $\frac{1}{2}m\omega^2 a^2 (\sin^2 \omega t + \cos^2 \omega t) = \frac{1}{2}m\omega^2 a^2$ = *constant*.

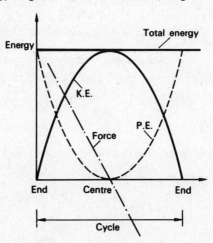

FIG. 2.18 Energy of S.H.M.

Thus the total energy of the vibrating mass and spring is constant. When the K.E. of the mass is a maximum (energy = $\frac{1}{2}m\omega^2 a^2$ and mass passing through the centre of oscillation), the P.E. of the spring

SIMPLE HARMONIC MOTION

is then zero ($x = 0$). Conversely, when the P.E. of the spring is a maximum (energy $= \frac{1}{2}ka^2 = \frac{1}{2}m\omega^2 a^2$ and mass at end of the oscillation), the K.E. of the mass is zero ($v = 0$). Fig. 2.18 shows the variation of P.E. and K.E. with displacement x over a cycle; the variation of the force F extending the spring, also shown, is directly proportional to the displacement from the centre of oscillation.

The constant interchange of energy between potential and kinetic energies is essential for producing and maintaining oscillations, whatever their nature. In the case of the oscillating bob of a simple pendulum, for example, the bob loses kinetic energy after passing through the middle of the swing, and then stores the energy as potential energy as it rises to the top of the swing. The reverse occurs as it swings back. In the case of oscillating layers of air when a sound wave passes, kinetic energy of the moving air molecules is converted to potential energy when the air is compressed. In the case of electrical oscillations, a coil L and a capacitor C in the circuit constantly exchange energy; this is stored alternately in the magnetic field of L and the electric field of C.

EXAMPLES

1. Define *simple harmonic motion* and state the relation between displacement from its mean position and the restoring force when a body executes simple harmonic motion.

A body is supported by a spiral spring and causes a stretch of 1·5 cm in the spring. If the mass is now set in vertical oscillation of small amplitude, what is the periodic time of the oscillation? (L.)

First part. Simple harmonic motion is the motion of an object whose acceleration is proportional to its distance from a fixed point and is always directed towards that point. The relation is: Restoring force $= -k \times$ distance from fixed point, where k is a constant.

Second part. Let m be the mass of the body in kg. Then, since 1·5 cm $= 0·015$ m

$$mg = k \times 0·015 \quad . \quad . \quad . \quad . \quad . \quad (i)$$

where k is a constant of the spring in N m^{-1}. Suppose the vibrating body is x m below its original position at some instant and is moving downwards. Then since the extension is $(x + 0·015)$ m, the net downward force

$$= mg - k(x + 0·015)$$
$$= mg - k \times 0·015 - kx = -kx$$

from (i). Now mass $\times$ acceleration = force.

$$\therefore m \times \text{acceleration} = -kx$$
$$\therefore \text{acceleration} = -\frac{k}{m}x.$$

But, from (i),
$$\frac{k}{m} = \frac{g}{0·015}$$

$$\therefore \text{acceleration} = \frac{g}{0.015}x = -\omega^2 x,$$

where $\omega^2 = g/0.015$.

$$\therefore \text{period } T = \frac{2\pi}{\omega} = 2\pi\sqrt{\frac{0.015}{g}} = 2\pi\sqrt{\frac{0.015}{9.8}}$$

$$= 0.25 \text{ second.}$$

2. A small bob of mass 20·00 g oscillates as a simple pendulum, with amplitude 5 cm and period 2 seconds. Find the velocity of the bob and the tension in the supporting thread, when the velocity of the bob is a maximum.

First part. See text.

Second part. The velocity, v, of the bob is a maximum when it passes through its original position. With the usual notation (see p. 51), the maximum velocity v_m is given by

$$v_m = \omega r,$$

where r is the amplitude of 0·05 m. Since $T = 2\pi/\omega$,

$$\therefore \omega = \frac{2\pi}{T} = \frac{2\pi}{2} = \pi \quad . \quad . \quad . \quad . \quad . \quad \text{(i)}$$

$$\therefore v_m = \omega r = \pi \times 0.05 = 0.16 \text{ m s}^{-1}.$$

Suppose P is the tension in the thread. The net force towards the centre of the circle along which the bob moves is then given by $(P - mg)$. The acceleration towards the centre of the circle, which is the point of suspension, is v_m^2/l, where l is the length of the pendulum.

$$\therefore P - mg = \frac{mv_m^2}{l}$$

$$\therefore P = mg + \frac{mv_m^2}{l}$$

Now
$$T = 2\pi\sqrt{\frac{l}{g}}$$

$$\therefore l = \frac{gT^2}{4\pi^2} = \frac{g \times 4}{4\pi^2} = \frac{g}{\pi^2}$$

Since $m = 0.02$ kg, $g = 9.8$ m s^{-2}, it follows from above that

$$P = 0.02 \times 9.8 + \frac{0.02 \times (0.05\pi)^2 \times \pi^2}{9.8}$$

$$= 19.65 \times 10^{-2} \text{ newton}$$

Waves. Wave equation

Waves and their properties can be demonstrated by producing them on the surface of water, as in a ripple tank. As the wave travels outwards from the centre of disturbance, it reaches more distant particles of water at a later time. Thus the particles of water vibrate out of *phase* with each other while the wave travels. It should be noted that the vibrating particles are the origin of the wave. Their mean position remains the same as the wave travels, but like the simple harmonic oscillators previously discussed, they store and release *energy* which is handed on from one part of the medium to another. The wave shows the energy travelling through the medium.

If the displacement y of a vibrating particle P is represented by $y = a \sin \omega t$, the displacement of a neighbouring particle Q can be represented by $y = a \sin(\omega t + \phi)$. ϕ is called the *phase angle* between the two vibrations. If $\phi = \pi/2$ or $90°$, the vibration of Q is $y = a \sin(\omega t + \pi/2)$. In this case, $y = 0$ when $t = 0$ for P, but $y = a \sin \pi/2 = a$ when $t = 0$ for Q. Comparing the two simple harmonic variations, it can be seen that Q *leads* on P by a quarter of a period.

If the wave is 'frozen' at different times, the displacements of the various particles will vary according to their position or distance x from some chosen origin such as the centre of disturbance. Now the *wavelength*, λ, of a wave is the distance between successive crests or troughs. At these points the phase difference is 2π. Consequently the phase angle for a distance x is $(x/\lambda) \times 2\pi$ or $2\pi x/\lambda$. The *wave equation*, which takes x into account as well as the time t, can thus be written as:

$$y = a \sin\left(2\pi\frac{t}{T} - 2\pi\frac{x}{\lambda}\right) = a \sin 2\pi\left(\frac{t}{T} - \frac{x}{\lambda}\right) \quad . \quad . \quad (1)$$

Other forms of the wave equation may be used. The velocity v of a wave is the distance travelled by the disturbance in 1 second. If the frequency of the oscillations is f, then f waves travel outwards in 1 second. Each wave occupies a length λ. Hence $v = f\lambda$. Further, the period T is the time for 1 oscillation. Thus $f = 1/T$ and hence $v = f\lambda = \lambda/T$. Substituting for T in (1), the wave equation may also be written as:

$$y = a \sin \frac{2\pi}{\lambda}(vt - x) \quad . \quad . \quad . \quad (2)$$

The wave equation in (1) or (2) is a *progressive wave*. The energy of the wave travels outwards through the medium as time goes on.

Longitudinal and transverse waves

Waves can be classified according to the direction of their vibrations. A *longitudinal wave* is one produced by *vibrations parallel* to the direction of travel of the wave. An example is a sound wave. The layers of air are always vibrating in a direction parallel to the direction of travel of the wave. A longitudinal wave can be seen travelling in a 'Slinky' coil when one end is fixed and the other is pulled to-and-fro in the direction of the coil.

A *transverse wave* is one produced by *vibrations perpendicular* to the direction of travel of the wave. Light waves are transverse waves. The wave along a bowed string of a violin is a transverse wave.

Velocity of waves

There are various types of waves. A longitudinal wave such as a sound wave is a *mechanical wave*. The speed v with which the energy travels depends on the restoring stress after particles in the medium are strained from their original position. Thus v depends on the *modulus of elasticity* of the medium. It also depends on the inertia of the particles, of which the mass per unit volume or density ρ is a measure.

By dimensions, as well as rigorously, it can be shown that

$$v = \sqrt{\frac{\text{modulus of elasticity}}{\rho}}.$$

For a solid, the modulus is Young's modulus, E. Thus $v = \sqrt{E/\rho}$. For a liquid or gas, the modulus is the bulk modulus, k. Hence $v = \sqrt{k/\rho}$. In air, $k = \gamma p$, where γ is the ratio of the principal specific heats of air and p is the atmospheric pressure. Thus $v = \sqrt{\gamma p/\rho}$ (p. 190).

When a taut string is plucked or bowed, the velocity of the transverse wave along it is given by $v = \sqrt{T/m}$, where T is the tension and m is the mass per unit length of the string. In this case T provides the restoring force acting on the displaced particles of string and m is a measure of their inertia.

Electromagnetic waves, which are due to electric and magnetic vibrations, form an important group of waves in nature. Radio waves, infra-red, visible and ultra-violet light, X-rays, and γ-rays are all electromagnetic waves, ranging from long wavelength such as 1000 metres (radio waves) to short wavelengths such as 10^{-11} m (γ-waves). Unlike the mechanical waves, no material medium is needed to carry the waves. The speed of all electromagnetic waves in a vacuum is the same, about 3×10^8 metre per second. The speed varies with wavelength in material media and this explains why dispersion (separation of colours) of white light is produced by glass.

Stationary waves

The equation $y = a \sin 2\pi(t/T - x/\lambda)$ represents a progressive wave travelling in the x-direction. A wave of the same amplitude and frequency travelling in the *opposite direction* is represented by the same form of equation but with $-x$ in place of x, that is, by $y = a \sin 2\pi(t/T + x/\lambda)$.

The *principle of superposition* states that the combined effect or resultant of two waves in a medium can be obtained by adding the displacements at each point due to the respective waves. Thus if the displacement due to one wave is represented by y_1, and that due to the other wave by y_2, the resultant displacement y is given by

$$y = y_1 + y_2 = a \sin 2\pi\left(\frac{t}{T} - \frac{x}{\lambda}\right) + a \sin 2\pi\left(\frac{t}{T} + \frac{x}{\lambda}\right)$$

$$= 2a \sin 2\pi\frac{t}{T} \cdot \cos 2\pi\frac{x}{\lambda} = A \sin 2\pi\frac{t}{T},$$

where $A = 2a \cos 2\pi x/\lambda$.

A represents the amplitude at different points in the medium. When $x = 0$, $y = A$; when $x = \lambda/4$, $A = 0$; when $x = \lambda/2$, $y = -A$; when $x = -3\lambda/4$, $y = 0$. Thus at some points called *antinodes*, A, the amplitude of vibration is a maximum. At points half-way between the antinodes called *nodes*, N, the amplitude is zero, that is, there is no vibration here. Fig. 2.19 (i). This type of wave, which stays in one place in a medium, is called a *stationary* or *standing wave*. Stationary waves may be produced which are either longitudinal or transverse.

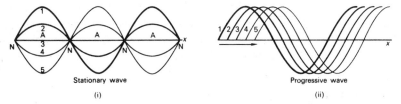

FIG. 2.19 Stationary and progressive waves

Unlike the progressive wave, where the energy travels outwards through the medium, Fig. 2.19 (ii), the energy of the stationary wave remains stored in one part of the medium. Stationary waves are produced in musical instruments when they are played. Stationary radio waves are also produced in receiving aerials. Stationary waves, due to electron motion, are believed to be present around the nucleus of atoms.

Interference. Diffraction

A stationary wave is a special case of *interference* between two waves. Another example occurs when two tuning forks of nearly equal frequency are sounded together. A periodic variation of loud sounds called 'beats' is then heard. They are due to the periodic variation of the amplitude of the resultant wave. If two very close coherent sources of light are obtained, interference between the two waves may produce bright and dark bands.

Diffraction is the name given to the interference between waves coming from coherent sources on the same undivided wavefront. The effect is pronounced when a wave is incident on a narrow opening whose width is of comparable order to the wavelength. The wave now spreads out or is 'diffracted' after passing through the slit. If the width of the slit, however, is large compared with the wavelength, the wave passes straight through the opening without any noticeable diffraction. This is why visible light, which has wavelengths of the order of only 10^{-7} m, passes straight through wide openings and produces sharp shadows; whereas sound, which has wavelengths over a million times longer and of the order of say 0·5 m, can be heard round corners.

Further details of wave phenomena are outside the scope of the work and may be obtained from *Light and Sound* by the author.

GRAVITATION

Kepler's Laws

The motion of the planets in the heavens had excited the interest of the earliest scientists, and Babylonian and Greek astronomers were able to predict their movements fairly accurately. It was considered for some time that the earth was the centre of the universe, but about 1542 COPERNICUS suggested that the planets revolved round the sun as centre. A great advance was made by KEPLER about 1609. He had studied for many years the records of observations on the planets made by TYCHO BRAHE, and he enunciated three laws known by his name. These state:

(1) The planets describe ellipses about the sun as one focus.

(2) The line joining the sun and the planet sweeps out equal areas in equal times.

(3) The squares of the periods of revolution of the planets are proportional to the cubes of their mean distances from the sun.

The third law was announced by Kepler in 1619.

Newton's Law of Gravitation

About 1666, at the early age of 24, NEWTON discovered a universal law known as the *law of gravitation*.

He was led to this discovery by considering the motion of a planet moving in a circle round the sun S as centre. Fig. 2.20 (i). The force acting on the planet of mass m is $mr\omega^2$, where r is the radius of the

GRAVITATION

circle and ω is the angular velocity of the motion (p. 42). Since $\omega = 2\pi/T$, where T is the period of the motion,

$$\text{force on planet} = mr\left(\frac{2\pi}{T}\right)^2 = \frac{4\pi^2 mr}{T^2}.$$

This is equal to the force of attraction of the sun on the planet. *Assuming an inverse-square law*, then, if k is a constant,

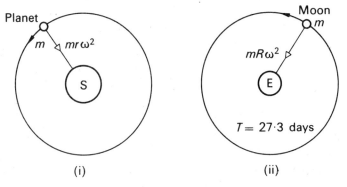

FIG. 2.20 Satellites

$$\text{force on planet} = \frac{km}{r^2}.$$

$$\therefore \frac{km}{r^2} = \frac{4\pi^2 mr}{T^2}$$

$$\therefore T^2 = \frac{4\pi^2}{k} r^3$$

$$\therefore T^2 \propto r^3,$$

since k, π are constants.

Now Kepler had announced that the squares of the periods of revolution of the planets are proportional to the cubes of their mean distances from the sun (see above). Newton thus suspected that *the force between the sun and the planet was inversely proportional to the square of the distance between them*. The great scientist now proceeded to test the inverse-square law by applying it to the case of the moon's motion round the earth. Fig. 2.20 (ii). The moon has a period of revolution, T, about the earth of approximately 27·3 days, and the force on it $= mR\omega^2$, where R is the radius of the moon's orbit and m is its mass.

$$\therefore \text{force} = mR\left(\frac{2\pi}{T}\right)^2 = \frac{4\pi^2 mR}{T^2}.$$

If the planet were at the earth's surface, the force of attraction on it due to the earth would be mg, where g is the acceleration due to gravity. Fig. 2.20 (ii). Assuming that the force of attraction varies as the inverse square of the distance between the earth and the moon,

$$\therefore \frac{4\pi^2 mR}{T^2} : mg = \frac{1}{R^2} : \frac{1}{r^2},$$

where r is the radius of the earth.

$$\therefore \frac{4\pi^2 R}{T^2 g} = \frac{r^2}{R^2},$$

$$\therefore g = \frac{4\pi^2 R^3}{r^2 T^2}. \quad . \quad . \quad . \quad (1)$$

Newton substituted the then known values of R, r, and T, but was disappointed to find that the answer for g was not near to the observed value, $9\cdot 8$ m s^{-2}. Some years later, he heard of a new estimate of the radius of the moon's orbit, and on substituting its value he found that the result for g was close to $9\cdot 8$ m s^{-2}. Newton saw that a universal law could be formulated for the attraction between any two particles of matter. He suggested that: *The force of attraction between two given masses is inversely proportional to the square of their distance apart.*

Gravitational Constant, G, and its Determination

From Newton's law, it follows that the force of attraction, F, between two masses m, M at a distance r apart is given by $F \propto \dfrac{mM}{r^2}$.

$$\therefore F = G\frac{mM}{r^2}, \quad . \quad . \quad . \quad (2)$$

where G is a universal constant known as the *gravitational constant*. This expression for F is *Newton's law of gravitation*.

From (2), it follows that G can be expressed in 'N m^2 kg^{-2}'. The dimensions of G are given by

$$[G] = \frac{MLT^{-2} \times L^2}{M^2} = M^{-1}L^3 T^{-2}.$$

Thus the unit of G may also be expressed as m^3 kg^{-1} s^{-2}.

A celebrated experiment to measure G was carried out by C. V. BOYS in 1895, using a method similar to one of the earliest determinations of G by CAVENDISH in 1798. Two identical balls, a, b, of gold,

GRAVITATION

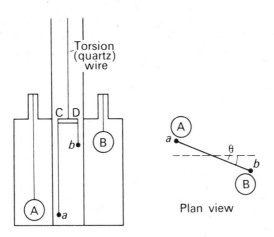

Fig. 2.21 Experiment on G

5 mm in diameter, were suspended by a long and a short fine quartz fibre respectively from the ends, C, D, of a highly-polished bar CD, Fig. 2.21. Two large identical lead spheres, A, B, 115 mm in diameter, were brought into position near a, b respectively. As a result of the attraction between the masses, two equal but opposite forces acted on CD. The bar was thus deflected, and the angle of deflection, θ, was measured by a lamp and scale method by light reflected from CD. The high sensitivity of the quartz fibres enabled the small deflection to be measured accurately, and the small size of the apparatus allowed it to be screened considerably from air convection currents.

Calculation for G

Suppose d is the distance between a, A, or b, B, when the deflection is θ. Then if m, M are the respective masses of a, A,

$$\text{torque of couple on CD} = G\frac{mM}{d^2} \times \text{CD}.$$

But torque of couple $= c\theta$,

where c is the torque in the torsion wire per unit radian of twist (p. 192).

$$\therefore G\frac{mM}{d^2} \times \text{CD} = c\theta.$$

$$\therefore G = \frac{c\theta d^2}{mM \times \text{CD}} \qquad . \qquad . \qquad . \qquad (1)$$

The constant c was determined by allowing CD to oscillate through a small angle and then observing its period of oscillation, T, which was of the order of 3 minutes. If I is the known moment of inertia of the system about the torsion wire, then (see p. 192),

$$T = 2\pi \sqrt{\frac{I}{c}}.$$

The constant c can now be calculated, and by substitution in (i), G can be determined. Accurate experiments showed that $G = 6.66 \times 10^{-11}$ N m² kg⁻² and HEYL in 1942, found G to be 6.67×10^{-11} N m² kg⁻².

Mass and Density of Earth

At the earth's surface the force of attraction on a mass m is mg, where g is the acceleration due to gravity. Now it can be shown that it is legitimate in calculations to assume that the mass, M, of the earth is concentrated at its centre, if it is a sphere. Assuming that the earth is spherical and of radius r, it then follows that the force of attraction of the earth on the mass m is GmM/r^2.

$$\therefore G\frac{mM}{r^2} = mg.$$

$$\therefore g = \frac{GM}{r^2}.$$

$$\therefore M = \frac{gr^2}{G}.$$

Now, $g = 9.8$ m s⁻², $r = 6.4 \times 10^6$ m, $G = 6.7 \times 10^{-11}$ N m² kg⁻².

$$\therefore M = \frac{9.8 \times (6.4 \times 10^6)^2}{6.7 \times 10^{-11}} = 6.0 \times 10^{24} \text{ kg}.$$

The volume of a sphere is $4\pi r^3/3$, where r is its radius. Thus the density, ρ, of the earth is approximately given by

$$\rho = \frac{M}{V} = \frac{gr^2}{4\pi r^3 G/3} = \frac{3g}{4\pi rG}.$$

By substituting known values of g, G, and r, the mean density of the earth is found to be about 5500 kg m⁻³. The density may approach a value of 10 000 kg m⁻³ towards the interior.

It is now believed that gravitational force travels with the speed of light. Thus if the gravitational force between the sun and earth were suddenly to disappear by the vanishing of the sun, it would take about 8 minutes for the effect to be experienced on the earth. The earth would then fly off along a tangent to its original curved path.

Gravitational and inertial mass

The mass m of an object appearing in the expression $F = ma$, force = mass × acceleration, is the *inertial mass*, as stated on p. 13. It is a measure of the reluctance of the object to move when forces act on it. It appears in $F = ma$ from Newton's second law of motion.

The 'mass' of the same object concerned in Newton's theory of gravitational attraction can be distinguished from the inertial mass. This is called the *gravitational mass*. If it is given the symbol m_g, then $F_g = GMm_g/r^2$, where F_g is the gravitational force, M is the mass of the earth and r its radius. Now $GM/r^2 = g$, the acceleration due to gravity (p. 70). Thus $F_g = m_g g = W$, the weight of the object.

In the simple pendulum theory on p. 55, we can derive the period T using $W = $ weight $= m_g g$ in place of the symbols adopted there.

Thus
$$-m_g g \frac{y}{l} = ma,$$

or
$$a = -\frac{m_g g}{ml} \cdot y = -\omega^2 y.$$

$$\therefore T = \frac{2\pi}{\omega} = 2\pi \sqrt{\frac{ml}{m_g g}}.$$

Experiments show that to a high degree of accuracy, $T = 2\pi\sqrt{l/g}$ no matter what mass is used, that is, the period depends only on l and g. Thus $m = m_g$, or the gravitational mass is equal to the inertial mass to the best of our present knowledge.

Mass of Sun

The mass M_S of the sun can be found from the period of a satellite and its distance from the sun. Consider the case of the earth. Its period T is about 365 days or $365 \times 24 \times 3600$ seconds. Its distance r_S from the centre of the sun is about $1 \cdot 5 \times 10^{11}$ m. If the mass of the earth is m, then, for circular motion round the sun,

$$\frac{GM_S m}{r_S^2} = mr_S \omega^2 = \frac{mr_S 4\pi^2}{T^2},$$

$$\therefore M_S = \frac{4\pi^2 r_S^3}{GT^2} = \frac{4\pi^2 \times (1 \cdot 5 \times 10^{11})^3}{6 \cdot 7 \times 10^{-11} \times (365 \times 24 \times 3600)^2} = 2 \times 10^{30} \text{ kg}.$$

Orbits round the earth

Satellites can be launched from the earth's surface to circle the earth. They are kept in their orbit by the gravitational attraction of the earth.

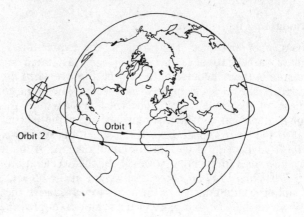

FIG. 2.22 Orbits round earth

Consider a satellite of mass m which just circles the earth of mass M close to its surface in an orbit 1. Fig. 2.22 (i). Then, if r is the radius of the earth,

$$\frac{mv^2}{r} = G\frac{Mm}{r^2} = mg,$$

where g is the acceleration due to gravity at the earth's surface and v is the velocity of m in its orbit. Thus $v^2 = rg$, and hence, using $r = 6.4 \times 10^6$ m and $g = 9.8$ m s^{-2},

$$v = \sqrt{rg} = \sqrt{6.4 \times 10^6 \times 9.8} = 8 \times 10^3 \text{ m s}^{-1} \text{ (approx)},$$

$$= 8 \text{ km s}^{-1}.$$

The velocity v in the orbit is thus about 8 km s^{-1}. In practice, the satellite is carried by a rocket to the height of the orbit and then given an impulse, by firing jets, to deflect it in a direction parallel to the tangent of the orbit (see p. 75). Its velocity is boosted to 8 km s^{-1} so that it stays in the orbit. The period in orbit

$$= \frac{\text{circumference of earth}}{v} = \frac{2\pi \times 6.4 \times 10^6 \text{ m}}{8 \times 10^3 \text{ m s}^{-1}}$$

$$= 5000 \text{ seconds (approx)} = 83 \text{ min}.$$

Parking Orbits

Consider now a satellite of mass m circling the earth in the plane of the equator in an orbit 2 concentric with the earth. Fig. 2.22 (ii). Suppose

GRAVITATION

the direction of rotation as the same as the earth and the orbit is at a distance R from the centre of the earth. Then if v is the velocity in orbit,

$$\frac{mv^2}{R} = \frac{GMm}{R^2}.$$

But $GM = gr^2$, where r is the radius of the earth.

$$\therefore \frac{mv^2}{R} = \frac{mgr^2}{R^2}$$

$$\therefore v^2 = \frac{gr^2}{R}.$$

If T is the period of the satellite in its orbit, then $v = 2\pi R/T$.

$$\therefore \frac{4\pi^2 R^2}{T^2} = \frac{gr^2}{R}$$

$$\therefore T^2 = \frac{4\pi^2 R^3}{gr^2}. \qquad\qquad\qquad\text{(i)}$$

If the period of the satellite in its orbit is exactly equal to the period of the earth as it turns about its axis, which is 24 hours, *the satellite will stay over the same place on the earth* while the earth rotates. This is sometimes called a 'parking orbit'. Relay satellites can be placed in parking orbits, so that television programmes can be transmitted continuously from one part of the world to another. *Syncom* was a satellite used for transmission of the Tokyo Olympic Games in 1964.

Since $T = 24$ hours, the radius R can be found from (i). Thus from

$$R = \sqrt[3]{\frac{T^2 gr^2}{4\pi^2}} \quad \text{and} \quad g = 9.8 \text{ m s}^{-2}, r = 6.4 \times 10^6 \text{ m,}$$

$$\therefore R = \sqrt[3]{\frac{(24 \times 3600)^2 \times 9.8 \times (6.4 \times 10^6)^2}{4\pi^2}} = 42{,}400 \text{ km}$$

The height above the earth's surface of the parking orbit

$$= R - r = 42\,400 - 6\,400 = 36\,000 \text{ km.}$$

In the orbit, the velocity of the satellite

$$= \frac{2\pi R}{T} = \frac{2\pi \times 42\,400}{24 \times 3600 \text{ seconds}} = 3.1 \text{ km s}^{-1}.$$

Weightlessness

When a rocket is fired to launch a spacecraft and astronaut into orbit round the earth, the initial acceleration must be very high owing to the large initial thrust required. This acceleration, a, is of the order of $15g$, where g is the gravitational acceleration at the earth's surface.

Suppose S is the reaction of the couch to which the astronaut is initially strapped. Fig. 2.23 (i). Then, from $F = ma$, $S - mg = ma = m \cdot 15g$, where m is the mass of the astronaut. Thus $S = 16mg$. This force is 16 times the weight of the astronaut and thus, initially, he experiences a large force.

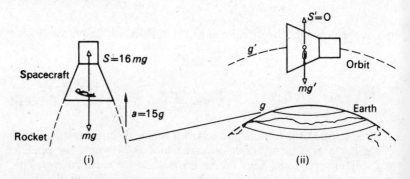

Fig. 2.23 Weight and weightlessness

In orbit, however, the state of affairs is different. This time the acceleration of the spacecraft and astronaut are both g' in magnitude, where g' is the acceleration due to gravity outside the spacecraft at the particular height of the orbit. Fig. 2.23 (ii). If S' is the reaction of the surface of the spacecraft in contact with the astronaut, then, for circular motion,

$$F = mg' - S' = ma = mg'.$$

Thus $S' = 0$. Consequently the astronaut becomes 'weightless'; he experiences no reaction at the floor when he walks about, for example. At the earth's surface we feel the reaction at the ground and are thus conscious of our weight. Inside a lift which is falling fast, the reaction at our feet diminishes. If the lift falls freely, the acceleration of objects inside is the same as that outside and hence the reaction on them is zero. This produces the sensation of 'weightlessness'. In orbit, as in Fig. 2.23 (ii), objects inside a spacecraft are also in 'free fall' because they have the same acceleration g' as the spacecraft. Consequently the sensation of weightlessness is experienced.

EXAMPLE

A satellite is to be put into orbit 500 km above the earth's surface. If its vertical velocity after launching is 2000 m s^{-1} at this height, calculate the magnitude and direction of the impulse required to put the satellite directly into orbit, if its mass is 50 kg. Assume $g = 10$ m s^{-2}; radius of earth, $R = 6,400$ km.

Suppose u is the velocity required for orbit, radius r. Then, with usual notation,

$$\frac{mu^2}{r} = \frac{GmM}{r^2} = \frac{gR^2m}{r^2}, \text{ as } \frac{GM}{R^2} = g.$$

$$\therefore u^2 = \frac{gR^2}{r}.$$

Now $R = 6,400$ km, $r = 6,900$ km, $g = 10$ m s^{-2}.

$$\therefore u^2 = \frac{10 \times (6,400 \times 10^3)^2}{6,900 \times 10^3}.$$

$$\therefore u = 7700 \text{ m s}^{-1} \text{ (approx.)}.$$

At this height, vertical momentum
$U_Y = mv = 50 \times 2,000 = 100,000$ kg m s^{-1}.
Fig. 2.24.

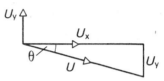

FIG. 2.24 Example

Horizontal momentum required $U_X = mu = 50 \times 7700 = 385,000$ kg m s^{-1}.

$$\therefore \text{impulse needed, } U, = \sqrt{U_Y^2 + U_X^2} = \sqrt{100,000^2 + 385,000^2}$$

$$= 4.0 \times 10^5 \text{ kg m s}^{-1} \quad . \quad . \quad . \quad . \quad (1)$$

Direction. The angle θ made by the total impulse with the horizontal or orbit tangent is given by $\tan \theta = U_Y/U_X = 100,000/385,000 = 0.260$. Thus $\theta = 14.6°$.

Magnitudes of acceleration due to gravity

(i) *Above the earth's surface.* Consider an object of mass m in an orbit of radius R from the centre, where $R > r$, the radius of the earth. Then, if g' is the acceleration due to gravity at this place,

$$mg' = \frac{GmM}{R^2} \quad . \quad . \quad . \quad . \quad (i)$$

But, if g is the acceleration due to gravity at the earth's surface,

$$mg = \frac{GmM}{r^2} \quad . \quad . \quad . \quad . \quad (ii)$$

Dividing (i) by (ii), $\therefore \dfrac{g'}{g} = \dfrac{r^2}{R^2}$, or $g' = \dfrac{r^2}{R^2} \cdot g$.

Thus above the earth's surface, the acceleration due to gravity g' varies *inversely as the square of the distance* from the centre. Fig. 2.25.

For a height h above the earth, $R = r+h$.

$$\therefore g' = \frac{r^2}{(r+h)^2} \cdot g = \frac{1}{\left(1+\frac{h}{r}\right)^2} \cdot g.$$

$$= \left(1+\frac{h}{r}\right)^{-2} \cdot g = \left(1-\frac{2h}{r}\right)g,$$

since powers of $(h/r)^2$ and higher can be neglected when h is small compared with r.

$$\therefore g - g' = \text{reduction in acceleration due to gravity.}$$

$$= \frac{2h}{r} \cdot g \quad . \quad . \quad . \quad . \quad . \quad . \quad . \quad (1)$$

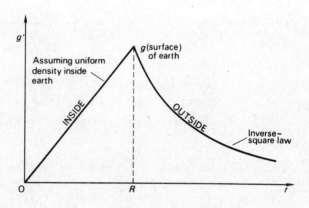

Fig. 2.25 Variation of g

(ii) *Below the earth's surface.* Consider an object of mass m at a point below the earth's surface. If its distance from the centre is b, the 'effective' mass M' of the earth which attracts it is that contained in a sphere of radius b. Assuming a constant density, then, since the mass of a sphere is proportional to *radius*^{3},

$$M' = \frac{b^3}{r^3} M,$$

where M is the mass of the earth. Suppose g'' is the acceleration due to gravity at the radius b. Then, from above,

$$mg'' = \frac{GmM'}{b^2} = \frac{GmMb}{r^3}.$$

GRAVITATION

Since $GM/r^2 = g$, it follows by substitution that

$$g'' = \frac{b}{r}g.$$

Thus assuming a uniform density of core, which is not the case in practice, the acceleration due to gravity g'' is directly proportional to the distance from the centre. Fig. 2.25.

If the depth below the earth's surface is h, then $b = r - h$.

$$\therefore g'' = \left(\frac{r-h}{r}\right)g = \left(1 - \frac{h}{r}\right)g$$

$$\therefore g - g'' = \frac{h}{r}g \quad . \quad . \quad . \quad . \quad . \quad (2)$$

Comparing (1) and (2), it can be seen that the acceleration at a distance h below the earth's surface is *greater* than at the same distance h above the earth's surface.

Simple harmonic motion due to gravitation

(i) *Along diameter of earth.* Suppose a body of mass m is imagined thrown into the earth along a tunnel passing through its centre. At a point distant x from the centre, the force of attraction, P, is given by $F = GmM'/x^2$, where M' is the 'effective mass' of the earth. Hence, since $M' = x^3M/r^3$, where M is the mass of the earth and r its radius,

$$P = \frac{GmM'}{x^3} = \frac{GmMx}{r^3}.$$

Now force, F, = mass × acceleration = m × acceleration. Since the force and x are in opposite directions, it follows that

$$m \times \text{acceleration} = -\frac{GmM}{r^3}x.$$

$$\therefore \text{acceleration} = -\frac{GM}{r^3}x.$$

This relationship shows that the body oscillates with simple harmonic motion about the centre of the earth. Further, the period of oscillation T is given, since $GM/r^2 = g$, by

$$T = \frac{2\pi}{\omega} = 2\pi\sqrt{\frac{r^3}{GM}} = 2\pi\sqrt{\frac{r}{g}} \quad . \quad . \quad . \quad (1)$$

(ii) *Along chord of earth.* Suppose now that a body is thrown into the earth along a tunnel AB which is a 'chord' of the earth's circle. Fig. 2.26. At a point D distant b from the centre C, the force F of attraction towards C is given, from p. 77, by

$$F = \frac{GmM'}{b^2} = \frac{GmMb}{r^3}.$$

∴ force along AB, P, =

$$F \cos \theta = \frac{GmMb \cos \theta}{r^3}.$$

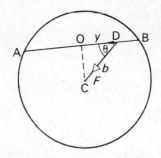

FIG. 2.26 S.H.M. along chord

But $b \cos \theta = y$ where OD $= y$ and O is the mid-point of AB.

$$\therefore P = \frac{GmMy}{r^3}.$$

∴ $m \times$ acceleration towards O $= -\dfrac{GmM}{r^3} . y.$

∴ acceleration towards O $= -\dfrac{GM}{r^3} . y.$

∴ motion is S.H.M. about O, and the period T is given by

$$T = \frac{2\pi}{\omega} = 2\pi \sqrt{\frac{r^3}{GM}} = \sqrt{\frac{r}{g}} \qquad . \qquad . \qquad . \quad (2)$$

The period of oscillation is thus the same along the chord and a diameter, from equation (1).

Potential

The *potential*, V, at a point due to the gravitational field of the earth is defined as numerically equal to the work done in taking a unit mass from infinity to that point. This is analogous to 'electric potential'. The potential at infinity is conventionally taken as *zero*.

For a point outside the earth, assumed spherical, we can imagine the whole mass M of the earth concentrated at its centre. The force of attraction on a unit mass outside the earth is thus GM/r^2, where r is the distance from the centre. The work done by the gravitational force in moving a distance δr towards the earth = force × distance = $GM . \delta r/r^2$. Hence the potential at a point distant a from the centre is given by

$$V_a = \int_{\infty}^{a} \frac{GM}{r^2} dr = -\frac{GM}{a} \qquad . \qquad . \qquad . \quad (1)$$

if the potential at infinity is taken as zero by convention. The negative sign indicates that the potential at infinity (zero) is *higher* than the potential close to the earth.

GRAVITATION

On the earth's surface, of radius r, we therefore obtain

$$V = -\frac{GM}{r} \qquad (2)$$

Velocity of Escape. Suppose a rocket of mass m is fired from the earth's surface Q so that it just escapes from the gravitational influence of the earth. Then work done = $m \times$ potential difference between infinity and Q.

$$= m \times \frac{GM}{r}.$$

$\therefore$ kinetic energy of rocket $= \frac{1}{2}mv^2 = m \times \dfrac{GM}{r}$.

$$\therefore v = \sqrt{\frac{2GM}{r}} = \text{velocity of escape.}$$

Now $\quad GM/r^2 = g.$

$$\therefore v = \sqrt{2gr}.$$

$\therefore v = \sqrt{2 \times 9 \cdot 8 \times 6 \cdot 4 \times 10^6} = 11 \times 10^3 \text{ m s}^{-1} = 11 \text{ km s}^{-1}$ (approx).

With an initial velocity, then, of about 11 km s^{-1}, a rocket will completely escape from the gravitational attraction of the earth. It can be made to travel towards the moon, for example, so that eventually it comes under the gravitational attraction of this planet. At present, 'soft' landings on the moon have been made by firing retarding retro rockets.

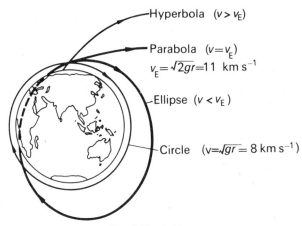

FIG. 2.27   Orbits

Summarising, with a velocity of about 8 km s^{-1}, a satellite can describe a circular orbit close to the earth's surface (p. 72). With a velocity greater than 8 km s^{-1} but less than 11 km s^{-1}, a satellite describes an elliptical orbit round the earth. Its maximum and minimum height in the orbit depends on its particular velocity. Fig. 2.27 illustrates the possible orbits of a satellite launched from the earth.

The molecules of air at normal temperatures and pressures have an average velocity of the order of 480 m s^{-1} or 0·48 km s^{-1} which is much less than the velocity of escape. Many molecules move with higher velocity than 0·48 km s^{-1} but gravitational attraction keeps the atmosphere round the earth. The gravitational attraction of the moon is much less than that of the earth and this accounts for the lack of atmosphere round the moon.

EXERCISES 2

(*Assume $g = 10\ m\ s^{-2}$*)

What are the missing words in the statements 1–6?

1. The force towards the centre in circular motion is called the ... force.

2. In simple harmonic motion, the maximum kinetic energy occurs at the ... of the oscillation.

3. The constant of gravitation G is related to g by ...

4. In simple harmonic motion, the maximum potential energy occurs at the ... of the oscillation.

5. Outside the earth, the acceleration due to gravity is proportional to ... from the centre.

6. A satellite in orbit in an equatorial plane round the earth will stay at the same place above the earth if its period is ... hours.

Which of the following answers, A, B, C, D or E, do you consider is the correct one in the statements 7–10?

7. The earth retains its atmosphere because *A* the earth is spherical, *B* the velocity of escape is greater than the mean speed of molecules, *C* the constant of gravitation is a universal constant, *D* the velocity of escape is less than the mean speed of molecules, *E* gases are lighter than solids.

8. In simple harmonic motion, the moving object has *A* only kinetic energy, *B* mean kinetic energy greater than the mean potential energy, *C* total energy equal to the sum of the maximum kinetic energy and maximum potential energy, *D* mean kinetic energy equal to the mean potential energy, *E* minimum potential energy at the centre of oscillation.

9. If r is the radius of the earth and g is the acceleration at its surface, then the acceleration g' at an orbit distance R from the centre of the earth is given by *A* $g'/g = r/R$, *B* $g'/g = r^2/R^2$, *C* $g'/g = R^2/r^2$, *D* $g'/g = (R-r)^2/r^2$, *E* $g'/g = (R-r)/r$.

10. When water in a bucket is whirled fast overhead, the water does not fall out at the top of the motion because *A* the centripetal force on the water is greater than the weight of water, *B* the force on the water is opposite to gravity, *C* the reaction of the bucket on the water is zero, *D* the centripetal force on the water is less than the weight of water, *E* atmospheric pressure counteracts the weight.

Circular Motion

11. An object of mass 4 kg moves round a circle of radius 6 m with a constant speed of 12 m s^{-1}. Calculate (i) the angular velocity, (ii) the force towards the centre.

12. An object of mass 10 kg is whirled round a horizontal circle of radius 4 m by a revolving string inclined to the vertical. If the uniform speed of the object is 5 m s^{-1}, calculate (i) the tension in the string, (ii) the angle of inclination of the string to the vertical.

13. A racing-car of 1000 kg moves round a banked track at a constant speed of 108 km h^{-1}. Assuming the total reaction at the wheels is normal to the track, and the horizontal radius of the track is 100 m, calculate the angle of inclination of the track to the horizontal and the reaction at the wheels.

14. An object of mass 8·0 kg is whirled round in a vertical circle of radius 2 m with a constant speed of 6 m s^{-1}. Calculate the maximum and minimum tensions in the string.

15. Define the terms (*a*) *acceleration*, and (*b*) *force*. Show that the acceleration of a body moving in a circular path of radius *r* with uniform speed *v* is v^2/r, and draw a diagram to show the direction of the acceleration.

A small body of mass *m* is attached to one end of a light inelastic string of length *l*. The other end of the string is fixed. The string is initially held taut and horizontal, and the body is then released. Find the values of the following quantities when the string reaches the vertical position: (*a*) the kinetic energy of the body, (*b*) the velocity of the body, (*c*) the acceleration of the body, and (*d*) the tension in the string. (*O. & C.*)

16. Explain what is meant by *angular velocity*. Derive an expression for the force required to make a particle of mass *m* move in a circle of radius *r* with uniform angular velocity *w*.

A stone of mass 500 g is attached to a string of length 50 cm which will break if the tension in it exceeds 20 N. The stone is whirled in a vertical circle, the axis of rotation being at a height of 100 cm above the ground. The angular speed is very slowly increased until the string breaks. In what position is this break most likely to occur, and at what angular speed? Where will the stone hit the ground? (*C.*)

Simple Harmonic Motion

17. An object moving with simple harmonic motion has an amplitude of 2 cm and a frequency of 20 Hz. Calculate (i) the period of oscillation, (ii) the acceleration at the middle and end of an oscillation, (iii) the velocities at the corresponding instants.

18. Calculate the length in centimetres of a simple pendulum which has a period of 2 seconds. If the amplitude of swing is 2 cm, calculate the velocity and acceleration of the bob (i) at the end of a swing, (ii) at the middle, (iii) 1 cm from the centre of oscillation.

19. Define *simple harmonic motion*. An elastic string is extended 1 cm when a small weight is attached at the lower end. If the weight is pulled down $\frac{1}{4}$ cm and then released, show that it moves with simple harmonic motion, and find the period.

20. A uniform wooden rod floats upright in water with a length of 30 cm immersed. If the rod is depressed slightly and then released, prove that its motion is simple harmonic and calculate the period.

21. A simple pendulum, has a period of 4·2 seconds. When the pendulum is shortened by 1 m, the period is 3·7 seconds. From these measurements, calculate the acceleration due to gravity and the original length of the pendulum.

22. What is *simple harmonic motion*? Show how it is related to the uniform motion of a particle with velocity v in a circle of radius r.

A steel strip, clamped at one end, vibrates with a frequency of 50 Hz and an amplitude of 8 mm at the free end. Find (*a*) the velocity of the end when passing through the zero position, (*b*) the acceleration at the maximum displacement.

23. Explain what is meant by *simple harmonic motion*.

Show that the vertical oscillations of a mass suspended by a light helical spring are simple harmonic and describe an experiment with the spring to determine the acceleration due to gravity.

A small mass rests on a horizontal platform which vibrates vertically in simple harmonic motion with a period of 0·50 second. Find the maximum amplitude of the motion which will allow the mass to remain in contact with the platform throughout the motion. (*L.*)

24. *Define* simple harmonic motion and *state* a formula for its period. Show that under suitable conditions the motion of a simple pendulum is simple harmonic and hence obtain an expression for its period.

If a pendulum bob is suspended from an inaccessible point, by a string whose length may be varied, describe how to determine (*a*) the acceleration due to gravity, (*b*) the height of the point of suspension above the floor.

How and why does the value of the acceleration due to gravity at the poles differ from its value at the equator? (*L.*)

25. Derive an expression for the time period of vertical oscillations of small amplitude of a mass suspended from the free end of a light helical spring.

What deformation of the wire of the spring occurs when the mass moves? (*N.*)

26. Give *two* practical examples of oscillatory motion which approximate to simple harmonic motion. What conditions must be satisfied if the approximations are to be good ones.

A point mass moves with simple harmonic motion. Draw on the same axes sketch graphs to show the variation with position of (*a*) the potential energy, (*b*) the kinetic energy, and (*c*) the total energy of the particle.

A particle rests on a horizontal platform which is moving vertically in simple harmonic motion with an amplitude of 10 cm. Above a certain frequency, the thrust between the particle and the platform would become zero at some point in the motion. What is this frequency, and at what point in the motion does the thrust become zero at this frequency? (*C.*)

27. In what circumstances will a particle execute simple harmonic motion? Show how simple harmonic motion can be considered to be the projection on the diameter of a circle of the motion of a particle describing the circle with uniform speed.

The balance wheel of a watch vibrates with an angular amplitude of π radians and a period of 0·5 second. Calculate (*a*) the maximum angular speed, (*b*) the angular speed when the displacement is $\pi/2$, and (*c*) the angular acceleration when the displacement is $\pi/4$. If the radius of the wheel is r, calculate the maximum radial force acting on a small dust particle of mass m situated on the rim of the wheel. (*O. & C.*)

28. Prove that the bob of a simple pendulum may move with simple harmonic motion, and find an expression for its period.

Describe with full details how you would perform an experiment, based on the expression derived, to measure the value of the acceleration due to gravity. What factors would influence your choice of (*a*) the length of the pendulum, (*b*) the material of the bob, and (*c*) the number of swings to be timed? (*O. & C.*)

29. Define *simple harmonic motion* and show that the free oscillations of a simple pendulum are simple harmonic for small amplitudes.

Explain what is meant by damping of oscillations and describe an experiment to illustrate the effects of damping on the motion of a simple pendulum. Briefly discuss the difficulties you would encounter and indicate qualitatively the results you would expect to observe. (*O. & C.*)

30. What is meant by simple harmonic motion? Obtain an expression for the kinetic energy of a body of mass m, which is performing S.H.M. of amplitude a and period $2\pi/\omega$, when its displacement from the origin is x.

Describe an experiment, or experiments, to verify that a mass oscillating at the end of a helical spring moves with simple harmonic motion. (*C.*)

31. State the dynamical condition under which a particle will describe simple harmonic motion. Show that it is approximately fulfilled in the case of the bob of a simple pendulum, and derive, from first principles, an expression for the period of the pendulum.

Explain how it can be demonstrated from observations on simple pendulums, that the weight of a body at a given place is proportional to its mass. (*O. & C.*)

32. Define *simple harmonic motion*. Show that a heavy body supported by a light spiral spring executes simple harmonic motion when displaced vertically from its equilibrium position by an amount which does not exceed a certain value and then released. How would you determine experimentally the maximum amplitude for simple harmonic motion?

A spiral spring gives a displacement of 5 cm for a load of 500 g. Find the maximum displacement produced when a mass of 80 g is dropped from a height of 10 cm on to a light pan attached to the spring. (*N.*)

Gravitation

33. Calculate the force of attraction between two small objects of mass 5 and 8 kg respectively which are 10 cm apart. ($G = 6·7 \times 10^{-11}$ N m^2 kg^{-2}.)

84 MECHANICS AND PROPERTIES OF MATTER

34. If the acceleration due to gravity is 9.8 m s^{-2} and the radius of the earth is 6400 km, calculate a value for the mass of the earth. ($G = 6.7 \times 10^{-11}$ N m^2 kg^{-2}.) Give the theory.

35. Assuming that the mean density of the earth is 5500 kg m^{-3}, that the constant of gravitation is 6.7×10^{-11} N m^2 kg^{-2}, and that the radius of the earth is 6400 km, find a value for the acceleration due to gravity at the earth's surface. Derive the formula used.

36. How do you account for the sensation of 'weightlessness' experienced by the occupant of a space capsule (*a*) in a circular orbit round the earth, (*b*) in outer space? Give one other instance in which an object would be 'weightless'. (*N*.)

37. State Newton's law of universal gravitation. Distinguish between the gravitational constant (*G*) and the acceleration due to gravity (*g*) and show the relation between them.
Describe an experiment by which the value of *g* may be determined. Indicate the measurements taken and how to calculate the result. Derive any formula used. (*L*.)

38. State *Newton's law of gravitation*. What experimental evidence is there for the validity of this law?
A binary star consists of two dense spherical masses of 10^{30} kg and 2×10^{30} kg whose centres are 10^7 km apart and which rotate together with a uniform angular velocity ω about an axis which intersects the line joining their centres. Assuming that the only forces acting on the stars arise from their mutual gravitational attraction and that each mass may be taken to act at its centre, show that the axis of rotation passes through the centre of mass of the system and find the value of ω. ($G = 6.7 \times 10^{-11}$ m^3 kg^{-1} s^{-2}.) (*O. & C.*)

39. Assuming that the planets are moving in circular orbits, apply Kepler's laws to show that the acceleration of a planet is inversely proportional to the square of its distance from the sun. Explain the significance of this and show clearly how it leads to Newton's law of universal gravitation.
Obtain the value of *g* from the motion of the moon, assuming that its period of rotation round the earth is 27 days 8 hours and that the radius of its orbit is 60·1 times the radius of the earth. (Radius of earth $= 6.36 \times 10^6$ m.) (*N*.)

40. Explain what is meant by the *gravitation constant* (*G*), and describe an accurate laboratory method of measuring it. Give an outline of the theory of your method.
Assuming that the earth is a sphere of radius 6,370 km and that $G = 6.66 \times 10^{-11}$ N m^2 kg^{-2}, calculate the mean density of the earth. (*O. & C.*)

41. Assuming the earth to be perfectly spherical, give sketch graphs to show how (*a*) the acceleration due to gravity, (*b*) the gravitational potential due to the earth's mass, vary with distance from the surface of the earth for points external to it. If any other assumption has been made, state what it is.
Explain why, even if the earth were a perfect sphere, the period of oscillation of a simple pendulum at the poles would not be the same as at the equator. Still assuming the earth to be perfectly spherical, discuss whether the velocity required to project a body vertically upwards, so that it rises to a given height, depends on the position on the earth from which it is projected. (*C*.)

GRAVITATION

42. Give the theory of the determination of the gravitational constant G by the Cavendish-Boys type of experiment, and derive an expression for the deflection in terms of G and quantities that are measured. Hence explain the superiority of Boys' experiment over the earlier one of Cavendish.

The earth may be regarded as a spherically uniform core of average density ρ_1, surrounded by a thin uniform shell of thickness h and density ρ_2. If the value of the acceleration due to gravity is the same at the surface as at depth h find the ratio of ρ_1 to ρ_2. (*N.*)

43. Explain what is meant by the *constant of gravitation*. Describe a laboratory experiment to determine it, showing how the result is obtained from the observations.

A proposed communication satellite would revolve round the earth in a circular orbit in the equatorial plane, at a height of 35,880 km above the earth's surface. Find the period of revolution of the satellite in hours, and comment on the result. (Radius of earth = 6,370 km, mass of earth = $5 \cdot 98 \times 10^{24}$ kg, constant of gravitation = $6 \cdot 66 \times 10^{-11}$ N m^2 kg^{-2}.) (*N.*)

44. State Newton's Law of gravitation. In what units is the gravitational constant measured?

Obtain an expression for the velocity acquired by a particle if it fell freely to the earth's surface starting from rest at an infinitely distant point. Neglect air resistance.

It can be proved that a uniform spherical shell of matter exerts no force on a massive particle placed inside it. Assuming the earth to behave as a sphere of uniform density, and that the external field for a spherical shell is the same as for an equal mass placed at its centre, calculate the force on a particle at a distance D from the centre of the earth. Use your result to sketch a graph showing how the force on such a particle would vary as it moved from the centre of the earth to a point well above the earth's surface. How would the form of this graph be changed if the density of the earth increased uniformly toward its centre? (*O. & C.*)

45. Give an account of the torsion balance method of measuring Newton's constant of gravitation, G. Two pieces of apparatus A and B for measuring G by the torsion balance are made of similar materials, and are constructed so that the linear dimensions of all parts of A, except the torsion wires, are n times as great as the corresponding parts of B. The torsion wires are so chosen that the two suspended systems have equal periods. Compare the deflections of the torsion bars of A and B.

Assuming that the moon describes a circular orbit of radius R about the earth in 27 days, and that Titan describes a circular orbit of radius $3 \cdot 2\,R$ about Saturn in 16 days, compare the masses of Saturn and the earth. (*O. & C.*)

Chapter 3

ROTATION OF RIGID BODIES

So far in this book we have considered the equations of motion and other dynamical formulae associated with a particle. In practice, however, an object is made of millions of particles, each at different places, and we need now to consider the motion of moving objects.

Moment of Inertia, I

Suppose a rigid object is rotating about a fixed axis O, and a particle A of the object makes an angle θ with a fixed line OY in space at some instant, Fig. 3.1. The angular velocity, $d\theta/dt$ or ω, of every particle about O is the same, since we are dealing with a rigid body, and the velocity v_1 of A at this instant is given by $r_1\omega$, where $r_1 =$ OA. Thus the kinetic energy of A $= \frac{1}{2}m_1v_1^2 = \frac{1}{2}m_1r_1^2\omega^2$. Similarly, the kinetic energy of another particle of the body $= \frac{1}{2}m_2r_2^2\omega^2$, where r_2 is its distance from O and m_2 is its mass. In this way we see that the kinetic energy, K.E., of the whole object is given by

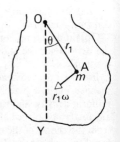

Fig. 3.1
Rotating rigid body

$$\text{K.E.} = \tfrac{1}{2}m_1r_1^2\omega^2 + \tfrac{1}{2}m_2r_2^2\omega^2 + \tfrac{1}{2}m_3r_3^2\omega^2 + \ldots$$
$$= \tfrac{1}{2}\omega^2(m_1r_1^2 + m_2r_2^2 + m_3r_3^2 + \ldots)$$
$$= \tfrac{1}{2}\omega^2(\Sigma mr^2),$$

where Σmr^2 represents the sum of the magnitudes of 'mr^2' for all the particles of the object. We shall see shortly how the quantity Σmr^2 can be calculated for a particular object. The magnitude of Σmr^2 is known as the *moment of inertia* of the object about the axis concerned, and we shall denote it by the symbol I. Thus

$$\text{Kinetic energy, K.E.,} = \tfrac{1}{2}I\omega^2. \qquad . \qquad . \qquad (1)$$

The units of I are *kg metre*^{2} (kg m^2). The unit of ω is 'radian s^{-1}' (rad s^{-1}). Thus if $I = 2$ kg m^2 and $\omega = 3$ rad s^{-1}, then

$$\text{K.E.} = \tfrac{1}{2}I\omega^2 = \tfrac{1}{2} \times 2 \times 3^2 \text{ joule} = 9 \text{ J}.$$

The kinetic energy of a particle of mass m moving with a velocity v is $\frac{1}{2}mv^2$. It will thus be noted that the formula for the kinetic energy of a rotating object is similar to that of a moving particle, the mass m being replaced by the moment of inertia I and the velocity v being replaced by the angular velocity ω. As we shall require values of I, the moment of inertia of several objects about a particular axis will first be calculated.

Moment of Inertia of Uniform Rod

(1) *About axis through middle.* The moment of inertia of a small element δx about an axis PQ through its centre O perpendicular to the length $= \left(\dfrac{\delta x}{l}M\right)x^2$, where l is the length of the rod, M is its mass, and x is the distance of the small element from O, Fig. 3.2.

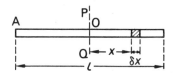

FIG. 3.2 Moment of inertia—uniform rod

$$\therefore \text{moment of inertia, } I, = 2\int_0^{l/2}\left(\frac{dx}{l}M\right)x^2$$

$$= \frac{2M}{l}\int_0^{l/2} x^2\,dx = \frac{Ml^2}{12} \quad . \quad (1)$$

Thus if the mass of the rod is 60 g and its length is 20 cm, $M = 6 \times 10^{-2}$ kg, $l = 0.2$ m, and $I = 6 \times 10^{-2} \times 0.2^2/12 = 2 \times 10^{-4}$ kg m².

(2) *About the axis through one end, A.* In this case, measuring distances x from A instead of O,

$$\text{moment of inertia, } I, = \int_0^l\left(\frac{dx}{l}M\right) \times x^2 = \frac{Ml^2}{3} \quad . \quad (2)$$

Moment of Inertia of Ring

Every element of the ring is the same distance from the centre. Hence the moment of inertia about an axis through the centre perpendicular to the plane of the ring $= Ma^2$, where M is the mass of the ring and a is its radius.

Moment of Inertia of Circular Disc

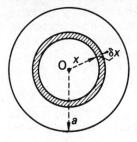

FIG. 3.3
Moment of inertia—disc

Consider the moment of inertia of a circular disc about an axis through its centre perpendicular to its plane, Fig. 3.3. If we take a small ring of the disc enclosed between radii x and $x + \delta x$, its mass $= \dfrac{2\pi x \delta x}{\pi a^2} M$, where a is the radius of the disc and M is its mass. Each element of the ring is distant x from the centre, and hence the moment of inertia of the ring about the axis through O $= \left(\dfrac{2\pi x \delta x}{\pi a^2} M\right) \times x^2$

$$\therefore \text{ moment of inertia of whole disc} = \int_0^a \dfrac{2\pi x \, dx}{\pi a^2} M \times x^2$$

$$= \dfrac{Ma^2}{2} \qquad . \qquad . \qquad (1)$$

Thus if the disc weighs 60 g and has a radius of 10 cm, $M = 60$ g $= 6 \times 10^{-2}$ kg, $a = 0\cdot1$ m, so that $I = 6 \times 10^{-2} \times 0\cdot1^2/2 = 3 \times 10^{-4}$ kg m^2.

Moment of Inertia of Cylinder

If a cylinder is *solid*, its moment of inertia about the axis of symmetry is the sum of the moments of inertia of discs into which we may imagine the cylinder cut. The moment of inertia of each disc $= \frac{1}{2}$ mass $\times a^2$, where a is the radius; and hence, if M is the mass of the cylinder,

$$\text{moment of inertia of solid cylinder} = \tfrac{1}{2}Ma^2 \qquad \text{(i)}$$

If a cylinder is *hollow*, its moment of inertia about the axis of symmetry is the sum of the moments of inertia of the curved surface and that of the two ends, assuming the cylinder is closed at both ends. Suppose a is the radius, h is the height of the cylinder, and σ is the mass per unit area of the surface. Then

$$\text{mass of curved surface} = 2\pi a h \sigma,$$

and   moment of inertia about axis $= $ mass $\times a^2 = 2\pi a^3 h \sigma$,

since we can imagine the surface cut into rings.

The moment of inertia of one end of the cylinder $= $ mass $\times a^2/2 = \pi a^2 \sigma \times a^2/2 = \pi a^4 \sigma/2$. Hence the moment of inertia of both ends $= \pi a^4 \sigma$.

$$\therefore \text{ moment inertia of cylinder}, I, = 2\pi a^3 h \sigma + \pi a^4 \sigma.$$

The mass of the cylinder, M, $= 2\pi ah\sigma + 2\pi a^2\sigma$

$$\therefore I = \frac{2\pi a^3 h\sigma + \pi a^4 \sigma}{2\pi ah\sigma + 2\pi a^2\sigma}M.$$

$$= \frac{2a^2h + a^3}{2h + 2a}M.$$

$$= \tfrac{1}{2}Ma^2 + \frac{a^2 h}{2h+2a}M \qquad . \qquad . \qquad . \qquad \text{(ii)}$$

If a hollow and a solid cylinder have the same mass M and the same radius and height, it can be seen from (i) and (ii) that the moment of inertia of the hollow cylinder is greater than that of the solid cylinder about the axis of symmetry. This is because the mass is distributed on the average at a greater distance from the axis in the former case.

Moment of Inertia of Sphere

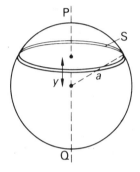

FIG. 3.4
Moment of inertia—sphere

The moment of inertia of a sphere about an axis PQ through its centre can be found by cutting thin discs such as S perpendicular to the axis, Fig. 3.4. The volume of the disc, of thickness δy and distance y from the centre,

$$= \pi r^2 \delta y = \pi(a^2 - y^2)\delta y.$$

$$\therefore \text{mass } M' \text{ of disc} = \frac{\pi(a^2 - y^2)\delta y}{4\pi a^3/3}M$$

$$= \frac{3M}{4a^3}(a^2 - y^2)\delta y,$$

where M is the mass of the sphere and a is its radius, since the volume of the sphere $= 4\pi a^3/3$. Now the moment of inertia of the disc about PO

$$= M' \times \frac{\text{radius}^2}{2} = \frac{3M}{4a^3}(a^2 - y^2)\delta y \times \frac{(a^2 - y^2)}{2}$$

$$\therefore \text{moment of inertia of sphere} = \frac{3M}{8a^3}\int_{-a}^{+a}(a^4 - 2a^2y^2 + y^4)dy$$

$$= \tfrac{2}{5}Ma^2 \qquad . \qquad . \qquad . \qquad (1)$$

Thus if the sphere weighs 4 kg and has a radius of 0·2 m, the moment of inertia $= \tfrac{2}{5} \times 4 \times 0·2^2 = 0·064$ kg m².

Radius of Gyration

The moment of inertia of an object about an axis, Σmr^2, is sometimes written as Mk^2, where M is the mass of the object and k is a quantity called the *radius of gyration* about the axis. For example, the moment of inertia of a rod about an axis through one end $= Ml^2/3$ (p. 87) $= M(l/\sqrt{3})^2$. Thus the radius of gyration, $k, = l/\sqrt{3} = 0.58l$. The moment of inertia of a sphere about its centre $= \frac{2}{5}Ma^2 = M \times (\sqrt{\frac{2}{5}}a)^2$. Thus the radius of gyration, $k, = \sqrt{\frac{2}{5}}a = 0.63a$ in this case.

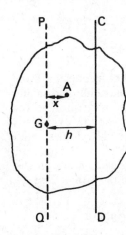

FIG. 3.5
Theorem of parallel axes

Relation Between Moment of Inertia About C.G. and Parallel Axis.

Suppose I is the moment of inertia of a body about an axis CD and I_G is the moment of inertia about a parallel axis PQ through the centre of gravity, G, distant h from the axis CD, Fig. 3.5. If A is a particle of mass m whose distance from PQ is x, its moment of inertia about CD $= m(h-x)^2$

$$\therefore I = \Sigma m(h-x)^2 = \Sigma mh^2 + \Sigma mx^2 - \Sigma 2mhx.$$

Now $\Sigma mh^2 = h^2 \times \Sigma m = Mh^2$, where M is the total mass of the object, and $\Sigma mx^2 = I_G$, the moment of inertia through the centre of gravity.

Also, $\Sigma 2mhx = 2h\Sigma mx = 0,$

since Σmx, the sum of the moments about the centre of gravity, is zero; this follows because the moment of the resultant (the weight) about G is zero.

$$\therefore I = I_G + Mh^2 \quad . \quad . \quad . \quad . \quad (1)$$

From this result, it follows that the moment of inertia, I, of a disc of radius a and mass M about an axis through a point on its circumference $= I_G + Ma^2$, since $h = a = $ radius of disc in this case. But $I_G = $ moment of inertia about the centre $= Ma^2/2$ (p. 88).

$$\therefore \text{moment of inertia}, I, = \frac{Ma^2}{2} + Ma^2 = \frac{3Ma^2}{2}.$$

Similarly the moment of inertia of a sphere of radius a and mass M about an axis through a point on its circumference $= I_G + Ma^2 = 2Ma^2/5 + Ma^2 = 7Ma^2/5$, since I_G, the moment of inertia about an axis through its centre, is $2Ma^2/5$.

Relation Between Moments of Inertia about Perpendicular Axes

Suppose OX, OY are any two perpendicular axes and OZ is an axis perpendicular to OX and OY, Fig. 3.6 (i). The moment of inertia, I, of

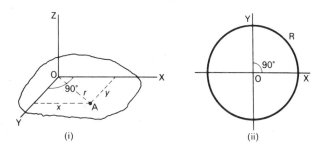

FIG. 3.6 Theorem of perpendicular axes

a body about the axis $OZ = \Sigma m r^2$, where r is the distance of a particle A from OZ and m is its mass. But $r^2 = x^2 + y^2$, where x, y are the distances of A from the axis OY, OX respectively.

$$\therefore I = \Sigma m(x^2 + y^2) = \Sigma m x^2 + \Sigma m y^2.$$

$$\therefore I = I_y + I_x, \qquad \qquad \qquad (1)$$

where I_y, I_x are the moments of inertia about OX, OY respectively.

As a simple application, consider a ring R and two perpendicular axes OX, OY in its plane, Fig. 3.6 (ii). Then from the above result,

$I_y + I_x = I =$ moment of inertia through O perpendicular to ring.

$$\therefore I_y + I_x = Ma^2.$$

But $I_y = I_x$, by symmetry.

$$\therefore I_x + I_x = Ma^2,$$

$$\therefore I_x = \frac{Ma^2}{2}.$$

This is the moment of inertia of the ring about any diameter in its plane.

In the same way, the moment of inertia, I, of a *disc* about a diameter in its plane is given by

$$I + I = \frac{Ma^2}{2},$$

since the moments of inertia, I, about the two perpendicular diameters are the same and $Ma^2/2$ is the moment of inertia of the disc about an axis perpendicular to its plane.

$$\therefore I = \frac{Ma^2}{4}.$$

Couple on a Rigid Body

Consider a rigid body rotating about a fixed axis O with an angular velocity ω at some instant. Fig. 3.7.

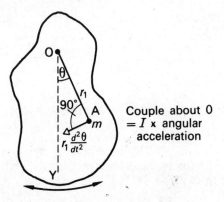

FIG. 3.7 Couple on rigid body

The force acting on the particle A = $m_1 \times$ acceleration = $m_1 \times \frac{d}{dt}(r_1\omega) = m_1 \times r_1 \frac{d\omega}{dt} = m_1 r_1 \frac{d^2\theta}{dt^2}$, since $\omega = \frac{d\theta}{dt}$. The moment of this force about the axis O = force × perpendicular distance from O = $m_1 r_1 \frac{d^2\theta}{dt^2} \times r_1$, since the force acts perpendicularly to the line OA.

$$\therefore \text{moment or } \textit{torque} = m_1 r_1^2 \frac{d^2\theta}{dt^2}.$$

$\therefore$ total moment of all forces on body about O, or *torque*,

$$= m_1 r_1^2 \frac{d^2\theta}{dt^2} + m_2 r_2^2 \frac{d^2\theta}{dt^2} + m_3 r_3^2 \frac{d^2\theta}{dt^2} + \ldots$$

$$= (\Sigma mr^2) \times \frac{d^2\theta}{dt^2},$$

since the angular acceleration, $d^2\theta/dt^2$, about O is the same for all particles.

$$\therefore \text{total torque about O } = I\frac{d^2\theta}{dt^2}, \quad . \quad . \quad (1)$$

ROTATION OF RIGID BODIES

where $I = \Sigma mr^2$ = moment of inertia about O. The moment about O is produced by external forces which together act as a *couple* of torque C say. Thus, for any rotating rigid body,

$$\text{Couple, } C = I\frac{d^2\theta}{dt^2}.$$

This result is analogous to the case of a particle of mass m which undergoes an acceleration a when a force F acts on it. Here $F = ma$. In place of F we have a *couple* C for a rotating rigid object; in place of m we have the *moment of inertia* I; and in place of linear acceleration a, we have *angular acceleration* $d^2\theta/dt^2 (d\omega/dt)$.

EXAMPLES

1. A heavy flywheel of mass 15 kg and radius 20 cm is mounted on a horizontal axle of radius 1 cm and negligible mass compared with the flywheel. Neglecting friction, find (i) the angular acceleration if a force of 40 N is applied tangentially to the axle, (ii) the angular velocity of the flywheel after 10 seconds.

(i) Moment of inertia $= \dfrac{Ma^2}{2} = \dfrac{15 \times 0.2^2}{2} = 0.3$ kg m².

Couple $C = 40 \text{ (N)} \times 0.01 \text{ (m)} = 0.4$ N m

∴ angular acceleration $= \dfrac{0.4}{0.3} = 1.3$ rad s^{-2}.

(ii) After 10 seconds, angular velocity = angular acceleration × time.

$= 1.3 \times 10 = 13$ rad s^{-1}.

2. The moment of inertia of a solid flywheel about its axis is 0·1 kg m². It is set in rotation by applying a tangential force of 20 N with a rope wound round the circumference, the radius of the wheel being 10 cm. Calculate the angular acceleration of the flywheel. What would be the acceleration if a mass of 2 kg were hung from the end of the rope? (*O. & C.*)

Couple $C = I\dfrac{d^2\theta}{dt^2}$ = moment of inertia × angular acceleration.

Now $C = 20 \times 0.1$ N m.

∴ angular acceleration $= \dfrac{20 \times 0.1}{0.1}$

$= 20$ rad s^{-2}.

If a mass of 2kg is hung from the end of the rope, it moves down with an acceleration a. Fig. 3.8. In this case, if T is the tension in the rope,

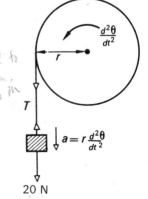

FIG. 3.8 Example

$$mg - T = ma \quad . \quad . \quad . \quad . \quad . \quad (1)$$

For the flywheel, $\quad T.r = \text{couple} = I\dfrac{d^2\theta}{dt^2} \quad . \quad . \quad . \quad (2)$

where r is the radius of the flywheel. Now the mass of 2 kg descends a distance given by $r\theta$, where θ is the angle the flywheel has turned. Hence the acceleration $a = rd^2\theta/dt^2$. Substituting in (1),

$$\therefore mg - T = mr\dfrac{d^2\theta}{dt^2}.$$

$$\therefore mgr - T.r = mr^2\dfrac{d^2\theta}{dt^2} \quad . \quad . \quad . \quad . \quad . \quad (3)$$

Adding (2) and (3),

$$\therefore mgr = (I + mr^2)\dfrac{d^2\theta}{dt^2}.$$

$$\therefore \dfrac{d^2\theta}{dt^2} = \dfrac{mgr}{I + mr^2} = \dfrac{2 \times 10 \times 0.1}{0.1 + 2 \times 0.1^2}.$$

$$= 16.7 \text{ rad s}^{-2},$$

using $g = 10$ m s^{-2}.

Angular Momentum and Conservation

In linear or straight-line motion, an important property of a moving object is its linear momentum (p. 20). When an object spins or rotates about an axis, its *angular momentum* plays an important part in its motion.

Consider a particle A of a rigid object rotating about an axis O. Fig. 3.9(i). The momentum of A = mass × velocity = $m_1 v = m_1 r_1 \omega$. The 'angular momentum' of A about O is defined as the *moment of the momentum* about O. Its magnitude is thus $m_1 v \times p$, where p is the perpendicular distance from O to the direction of v. Thus angular momentum of A = $m_1 v p = m_1 r_1 \omega \times r_1 = m_1 r_1^2 \omega$.

$$\therefore \text{total angular momentum of whole body} = \Sigma m_1 r_1^2 \omega = \omega \Sigma m_1 r_1^2$$

$$= I\omega,$$

where I is the moment of inertia of the body about O.

Angular momentum is analogous to 'linear momentum', mv, in the dynamics of a moving particle. In place of m we have I, the moment of inertia; in place of v we have ω, the angular velocity.

Further, the *conservation of angular momentum*, which corresponds to the conservation of linear momentum, states that *the angular momentum about an axis of a given rotating body or system of bodies is constant, if no external couple acts about that axis.* Thus when a high

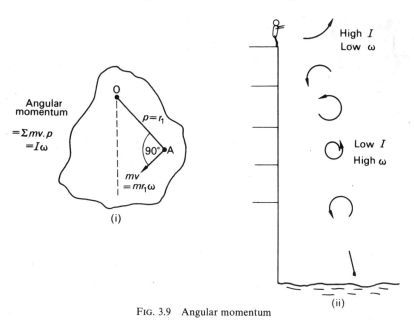

FIG. 3.9 Angular momentum

diver jumps from a diving board, his moment of inertia, I, can be decreased by curling his body more, in which case his angular velocity ω is increased. Fig. 3.9 (ii). He may then be able to turn more somersaults before striking the water. Similarly, a dancer on skates can spin faster by folding her arms.

The earth is an object which rotates about an axis passing through its geographic north and south poles with a period of 1 day. If it is struck by meteorites, then, since action and reaction are equal, no external couple acts on the earth and meteorites. Their total angular momentum is thus conserved. Neglecting the angular momentum of the meteorites about the earth's axis before collision compared with that of the earth, then

angular momentum of *earth plus meteorites* after collision = angular momentum of *earth* before collision.

Since the effective mass of the earth has increased after collision the moment of inertia has increased. Hence the earth will slow up slightly. Similarly, if a mass is dropped gently on to a turntable rotating freely at a steady speed, the conservation of angular momentum leads to a reduction in the speed of the table.

Angular momentum, and the principle of the conservation of angular momentum, have wide applications in physics. They are used in connection with enormous rotating masses such as the earth, as well as minute spinning particles such as electrons, neutrons and protons found inside atoms.

Experiment on Conservation of Angular Momentum

A simple experiment on the principle of the conservation of angular momentum is illustrated below.

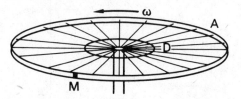

FIG. 3.10 Conservation of angular momentum

Briefly, in Fig. 3.10 (i) a bicycle wheel A without a tyre is set rotating in a horizontal plane and the time for three complete revolutions is obtained with the aid of a white tape marker M on the rim. A ring D of known moment of inertia, I, is then gently placed on the wheel concentric with it, by 'dropping' it from a small height. The time for the next three revolutions is then determined. This is repeated with several more rings of greater known moment of inertia.

If the principle of conservation of angular momentum is true, then $I_0\omega_0 = (I_0 + I_1)\omega_1$, where I_0 is the moment of inertia of the wheel alone, ω_0 is the angular frequency of the wheel alone, and ω_1 is the angular frequency with a ring. Thus if t_0, t_1 are the respective times for three revolutions,

$$\frac{I_0 + I_1}{t_1} = \frac{I_0}{t_0}$$

$$\therefore \frac{I_1}{I_0} + 1 = \frac{t_1}{t_0}.$$

Thus a graph of t_1/t_0 v. I_1 should be a straight line. Within the limits of experimental error, this is found to be the case.

EXAMPLE

Consider a disc of mass 100 g and radius 10 cm is rotating freely about axis O through its centre at 40 r.p.m. Fig. 3.11. Then, about O,

moment of inertia $I = \dfrac{Ma^2}{2} = \tfrac{1}{2} \times 0 \cdot 1 \text{ (kg)} \times 0 \cdot 1^2 \text{ (m}^2) = 5 \times 10^{-4}$ kg m^2,

and angular momentum $= I\omega = 5 \times 10^{-4}\omega$,

where ω is the angular velocity corresponding to 40 r.p.m.

Suppose some wax W of mass m 20 g is dropped gently on to the disc at a distance r of 8 cm from the centre O. The disc then slows down to another speed, corresponding to an angular velocity ω_1 say. The total angular momentum about O of disc plus wax

$$= I\omega_1 + mr^2\omega_1 = 5 \times 10^{-4}\omega_1 + 0.02 \times 0.08^2 \cdot \omega_1$$
$$= 6.28 \times 10^{-4}\omega_1.$$

From the conservation of angular momentum for the disc and wax about O

FIG. 3.11 Example

$$6.28 \times 10^{-4}\omega_1 = 5 \times 10^{-4}\omega.$$
$$\therefore \frac{\omega_1}{\omega} = \frac{500}{628} = \frac{n}{40},$$

where n is the r.p.m. of the disc.

$$\therefore n = \frac{500}{628} \times 40 = 32 \text{ (approx)}.$$

Kepler's law and angular momentum

Consider a planet moving in an orbit round the sun S. Fig. 3.12. At an instant when the planet is at O, its velocity v is along the tangent to the orbit at O. Suppose the planet moves a very small distance δs from O to B in a small time δt, so that the velocity $v = \delta s/\delta t$ and its direction is practically along OB. Then, if the conservation of angular momentum is obeyed,

$$mv \times p = \text{constant},$$

where m is the mass of the planet and p is the perpendicular from S to OB produced.

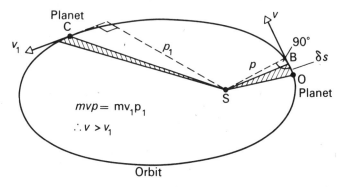

FIG. 3.12 Angular momentum and planets

$$\therefore \frac{m \cdot \delta s \cdot p}{\delta t} = \text{constant}.$$

But the area δA of the triangle SBO = $\frac{1}{2}$ base × height = $\delta s \times p/2$.

$$\therefore m \cdot 2\frac{\delta A}{\delta t} = \text{constant}$$

$$\therefore \frac{\delta A}{\delta t} = \text{constant},$$

since $2m$ is constant. Thus if the conservation of angular momentum is true, the area swept out per second by the radius SO is constant while the planet O moves in its orbit. In other words, equal areas are swept out in equal times. *But this is Kepler's second law*, which has been observed to be true for centuries (see p. 66). Consequently, the principle of the conservation of angular momentum has stood the test of time. From the equality of the angular momentum values at O and C, where p is less than p_1, it follows that v is greater than v_1. Thus the planet speeds up on approaching S.

The force on O is always one of attraction towards S. It is described as a *central force*. Thus the force has no moment about O and hence the angular momentum of the planet about S is conserved.

Kinetic Energy of a Rolling Object

When an object such as a cylinder or ball rolls on a plane, the object is rotating as well as moving bodily along the plane; therefore it has rotational energy as well as translational energy.

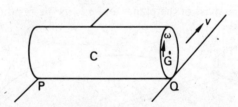

Fig. 3.13 Rolling object

Consider a cylinder C rolling along a plane without slipping, Fig. 3.13. At any instant the line of contact, PQ, with the plane is at rest, and we can consider the whole of the cylinder to be rotating about this axis. Hence the energy of the cylinder = $\frac{1}{2}I_1\omega^2$, where I_1 is the moment of inertia about PQ and ω is the angular velocity.

But if I is the moment of inertia about a parallel axis through the

ROTATION OF RIGID BODIES

centre of gravity of the cylinder, M is the mass of the cylinder and a its radius, then
$$I_1 = I + Ma^2,$$
from the result on p. 90.

$$\therefore \text{energy of cylinder} = \tfrac{1}{2}(I + Ma^2)\omega^2$$
$$= \tfrac{1}{2}I\omega^2 + \tfrac{1}{2}Ma^2\omega^2$$
$$\therefore \text{Energy} = \tfrac{1}{2}I\omega^2 + \tfrac{1}{2}Mv^2 \quad . \quad . \quad (38)$$

since, by considering the distance rolled and the angle then turned, $v = a\omega$ = velocity of centre of gravity. This energy formula is true for any moving object.

As an application of the energy formula, suppose a *ring* rolls along a plane. The moment of inertia about the centre of gravity, its centre, $= Ma^2$ (p. 87); also, the angular velocity, ω, about its centre $= v/a$, where v is the velocity of the centre of gravity.

$$\therefore \text{kinetic energy of ring} = \tfrac{1}{2}Mv^2 + \tfrac{1}{2}I\omega^2$$
$$= \tfrac{1}{2}Mv^2 + \tfrac{1}{2}Ma^2 \times \left(\frac{v}{a}\right)^2 = Mv^2.$$

By similar reasoning, the kinetic energy of a sphere rolling down a plane
$$= \tfrac{1}{2}Mv^2 + \tfrac{1}{2}I\omega^2$$
$$= \tfrac{1}{2}Mv^2 + \tfrac{1}{2} \times \tfrac{2}{5}Ma^2 \times \left(\frac{v}{a}\right)^2 = \tfrac{7}{10}Mv^2,$$
since $I = 2Ma^2/5$ (p. 89).

Acceleration of Rolling Object

We can now deduce the acceleration of a rolling object down an inclined plane.

As an illustration, suppose a solid cylinder rolls down a plane. Then
$$\text{kinetic energy} = \tfrac{1}{2}Mv^2 + \tfrac{1}{2}I\omega^2.$$
But moment of inertia, I, about an axis through the centre of gravity parallel to the plane $= \tfrac{1}{2}Ma^2$, and $\omega = v/a$, where a is the radius.
$$\therefore \text{kinetic energy} = \tfrac{1}{2}Mv^2 + \tfrac{1}{4}Mv^2 = \tfrac{3}{4}Mv^2.$$
If the cylinder rolls from *rest* through a distance s, the loss of potential energy $= Mgs \sin \alpha$, where α is the inclination of the plane to the horizontal.
$$\therefore \tfrac{3}{4}Mv^2 = Mgs \sin \alpha$$
$$\therefore v^2 = \frac{4g}{3}s \sin \alpha$$

But $v^2 = 2as$, where a is the linear acceleration.

$$\therefore 2as = \frac{4g}{3}s \sin \alpha$$

$$\therefore a = \frac{2g}{3} \sin \alpha \quad . \quad . \quad . \quad . \quad (i)$$

The acceleration if sliding, and no rolling, took place down the plane is $g \sin \alpha$. The cylinder has thus a smaller acceleration when rolling.

The time t taken to move through a distance s from rest is given by $s = \frac{1}{2}at^2$. Thus, from (i),

$$s = \tfrac{1}{3}gt^2 \sin \alpha,$$

$$\text{or} \quad t = \sqrt{\frac{3s}{g \sin \alpha}}.$$

If the cylinder is *hollow*, instead of solid as assumed, the moment of inertia about an axis through the centre of gravity parallel to the plane is greater than that for a solid cylinder, assuming the same mass and dimensions (p. 88). The time taken for a hollow cylinder to roll a given distance from rest on the plane is then greater than that taken by the solid cylinder, from reasoning similar to that above; and thus if no other means were available, a time test on an inclined plane will distinguish between a solid and a hollow cylinder of the same dimensions and mass. If a torsion wire is available, however, the cylinders can be suspended in turn, and the period of torsional oscillations determined. The cylinder of larger moment of inertia, the hollow cylinder, will have a greater period, as explained on p. 102.

Measurement of Moment of Inertia of Flywheel

The moment of inertia of a flywheel W about a horizontal axle A can be determined by tying one end of some string to a pin on the axle, winding the string round the axle, and attaching a mass M to the other end of the string, Fig. 3.14. The length of string is such that M reaches the floor, when released, at the same instant as the string is completely unwound from the axle.

M is released, and the number of revolutions, n, made by the wheel W up to the occasion when M strikes the ground is noted. The further number of revolutions n_1 made by W until it comes finally to rest, and the time t taken, are also observed by means of a chalk-mark on W.

Now the loss in potential energy of M = gain in kinetic energy of M + gain in kinetic energy of flywheel + work done against friction.

$$\therefore Mgh = \tfrac{1}{2}Mr^2\omega^2 + \tfrac{1}{2}I\omega^2 + nf, \quad . \quad . \quad (i)$$

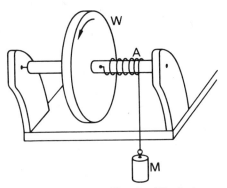

FIG. 3.14 Moment of inertia of flywheel

where h is the distance M has fallen, r is the radius of the axle, ω is the angular velocity, I is the moment of inertia, and f is the energy per turn expended against friction. Since the energy of rotation of the flywheel when the mass M reaches the ground = work done against friction in n_1 revolutions, then

$$\tfrac{1}{2}I\omega^2 = n_1 f.$$

$$\therefore f = \tfrac{1}{2}\frac{I\omega^2}{n_1}.$$

Substituting for f in (i),

$$\therefore Mgh = \tfrac{1}{2}Mr^2\omega^2 + \tfrac{1}{2}I\omega^2\left(1+\frac{n}{n_1}\right) \qquad . \qquad . \quad \text{(ii)}$$

Since the angular velocity of the wheel when M reaches the ground is ω, and the final angular velocity of the wheel is zero after a time t, the average angular velocity = $\omega/2 = 2\pi n_1/t$. Thus $\omega = 4\pi n_1/t$. Knowing ω and the magnitude of the other quantities in (ii), the moment of inertia I of the flywheel can be calculated.

Period of Oscillation of Rigid Body

On p. 92 we showed that the moment of the forces acting on rotating objects = $I d\omega/dt = I d^2\theta/dt^2$, where I is the moment of inertia about the axis concerned and $d^2\theta/dt^2$ is the angular acceleration about the axis. Consider a rigid body oscillating about a fixed axis O, Fig. 3.15. The moment of the weight mg (the only external force) about O is $mgh\sin\theta$, or $mgh\theta$ if θ is small, where h is the distance of the centre of gravity from O.

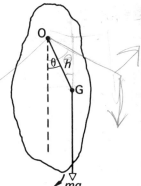

FIG. 3.15 Compound pendulum

$$\therefore I\frac{d^2\theta}{dt^2} = -mgh\theta,$$

the minus indicating that the moment due to the weight always *opposes* the growth of the angle θ.

$$\therefore \frac{d^2\theta}{dt^2} = \frac{-mgh}{I}\theta = -\omega^2\theta,$$

where $\omega^2 = mgh/I$.

∴ the motion is simple harmonic motion (p. 50),

and period, $T, = \dfrac{2\pi}{\omega} = \dfrac{2\pi}{\sqrt{mgh/I}} = 2\pi\sqrt{\dfrac{I}{mgh}}$ . . (1)

If $I = mk_1^2$, where k_1 is the radius of gyration about O,

$$T = 2\pi\sqrt{\frac{mk_1^2}{mgh}} = 2\pi\sqrt{\frac{k_1^2}{gh}} \qquad . \quad . \quad . \quad (2)$$

A *ring* of mass m and radius a will thus oscillate about an axis through a point O on its circumference normal to the plane of the ring with a period T given by

$$T = 2\pi\sqrt{\frac{I_0}{mga}}.$$

But $I_0 = I_G + ma^2$ (theorem of parallel axes) $= ma^2 + ma^2 = 2ma^2$.

$$\therefore T = 2\pi\sqrt{\frac{2ma^2}{mga}} = 2\pi\sqrt{\frac{2a}{g}}.$$

Thus if $a = 0.5$ m, $g = 9.8$ m s^{-2},

$$T = 2\pi\sqrt{\frac{2 \times 0.5}{9.8}} = 2.0 \text{ seconds (approx).}$$

Measurement of Moment of Inertia of Plate

The moment of inertia of a circular disc or other plate about an axis perpendicular to its plane, for example, can be measured by means of torsional oscillations. The plate is suspended horizontally from a vertical torsion wire, and the period T_1 of torsional oscillations is measured. Then, from (1),

$$T_1 = 2\pi\sqrt{\frac{I_1}{c}}, \qquad . \quad . \quad . \quad . \quad \text{(i)}$$

where I_1 is the moment of inertia and c is the constant (opposing couple per unit radian) of the wire (p. 192). A ring or annulus of *known* moment of inertia I_2 is now placed on the plate concentric with the axis, and the new period T_2 is observed. Then

ROTATION OF RIGID BODIES

$$T_2 = 2\pi \sqrt{\frac{I_1+I_2}{c}} . \quad . \quad . \quad . \quad \text{(ii)}$$

By squaring (i) and (ii), and then eliminating c, we obtain

$$I_1 = \frac{T_1^2}{T_2^2 - T_1^2} \cdot I_2.$$

Thus knowing T_1, T_2, and I_2, the moment of inertia I_1 can be calculated.

Compound Pendulum. Since $I = I_G + mh^2 = mk^2 + mh^2$, where I_G is the moment of inertia about the centre of gravity, h is the distance of the axis O from the centre of gravity, and k is the radius of gyration about the centre of gravity, then, from previous,

$$T = 2\pi \sqrt{\frac{I}{mgh}} = 2\pi \sqrt{\frac{mk^2 + mh^2}{mgh}}.$$

$$\therefore T = 2\pi \sqrt{\frac{k^2 + h^2}{gh}}.$$

Hence
$$T = 2\pi \sqrt{\frac{l}{g}},$$

where
$$l = \frac{k^2 + h^2}{h} \quad . \quad . \quad . \quad . \quad \text{(i)}$$

Thus $(k^2 + h^2)/h$ is the length, l, of the *equivalent simple pendulum*.

From (i), $\quad h^2 - hl + k^2 = 0.$

$$\therefore h_1 + h_2 = l, \text{ and } h_1 h_2 = k^2,$$

where h_1 and h_2 are the roots of the equation.

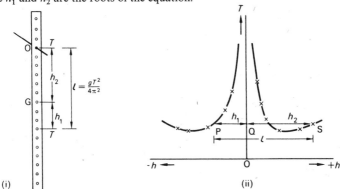

FIG. 3.16 Compound pendulum experiment

By timing the period of vibration, T, of a long rod about a series of axes at varying distances h on either side of the centre of gravity, and then plotting a graph of T v. h, two different values of h giving the same period can be obtained, Fig. 3.16 (i), (ii). Suppose h_1, h_2 are the two values. Then from the result just obtained, $h_1 + h_2 = l$, the length of the equivalent simple pendulum. Thus, since $T = 2\pi\sqrt{l/g}$,

$$g = \frac{4\pi^2 l}{T^2} = \frac{4\pi^2(h_1 + h_2)}{T^2}.$$

In Fig. 3.16 (ii), $PQ + QS = h_1 + h_2 = l$.

Kater's Pendulum. The acceleration due to gravity was first measured by the simple pendulum method, and calculated from the relation $g = 4\pi^2 l/T^2$, with the usual notation. The length l, the distance from the point of suspension to the centre of gravity of the bob, however, cannot be determined with very great accuracy.

In 1817 Captain Kater designed a reversible pendulum, with knife-edges for the suspension; it was a compound pendulum. Now it has just been shown that the same period is obtained between two non-symmetrical points on a compound pendulum when their distance apart is l, the length of the equivalent simple pendulum. Thus if T is the period about either knife-edge when this occurs, $g = 4\pi^2 l/T^2$, where l is now the distance between the knife-edges. The pendulum is made geometrically symmetrical about the mid-point, with a brass bob at one end and a wooden bob of the same size at the other. A movable large and small weight are placed between the knife-edges, which are about one metre apart. The period is then slightly greater than 2 seconds.

To find g, the pendulum is set up in front of an accurate seconds clock, with the bob of the clock and that of the Kater pendulum in line with each other, and both sighted through a telescope. The large weight on the pendulum is moved until the period is nearly the same about either knife-edge, and the small weight is used as a fine adjustment. When the periods are the same, the distance l between the knife-edges is measured very accurately by a comparator method with a microscope and standard metre. Thus knowing T and l, g can be calculated from $g = 4\pi^2 l/T^2$.

For details of the experiment the reader should consult *Advanced Practical Physics for Students* by Worsnop and Flint (Methuen).

Summary

The following table compares the translational (linear) motion of a small mass m with the rotational motion of a large object of moment of inertia I.

	Linear Motion	Rotational Motion
1.	Velocity, v	Velocity $v = r\omega$
2.	Momentum $= mv$	Angular momentum $= I\omega$
3.	Energy $= \frac{1}{2}mv^2$	Rotational energy $= \frac{1}{2}I\omega^2$
4.	Force, $F, = ma$	Torque, $C, = I \times$ ang. accn. $(d^2\theta/dt^2)$
5.	Simple pendulum: $T = 2\pi\sqrt{\dfrac{l}{g}}$	Compound pendulum: $T = 2\pi\sqrt{\dfrac{I}{mgh}}$
6.	Motion down inclined plane — energy equation: $\frac{1}{2}mv^2 = mgh\sin\theta$	Rotating without slipping down inclined plane — energy equation: $\frac{1}{2}Mv^2 + \frac{1}{2}I\omega^2 = Mgh\sin\theta$
7.	Conservation of linear momentum on collision, if no external forces	Conservation of angular momentum on collision, if no external couple

EXAMPLES

1. What is meant by the *moment of inertia* of an object about an axis?
Describe and give the theory of an experiment to determine the moment of inertia of a flywheel mounted on a horizontal axle.

A uniform circular disc of mass 20 kg and radius 15 cm is mounted on a horizontal cylindrical axle of radius 1·5 cm and negligible mass. Neglecting frictional losses in the bearings, calculate (a) the angular velocity acquired from rest by the application for 12 seconds of a force 20 N tangential to the axle, (b) the kinetic energy of the disc at the end of this period, (c) the time required to bring the disc to rest if a braking force of 1 N were applied tangentially to its rim. (*L.*)

Moment of inertia of disc, $I, = \frac{1}{2}Ma^2 = \frac{1}{2} \times 20 \, (\text{kg}) \times 0.15^2 \, (\text{m}^2) = 0.225 \text{ kg m}^2$.

(a) Torque due to 20 N tangential to axle
$$= 20 \, (\text{N}) \times 0.015 \, (\text{m}) = 0.3 \text{ N m}.$$

$\therefore$ angular acceleration $= \dfrac{\text{torque}}{I} = \dfrac{0.3}{0.225} \text{ rad s}^{-2}$.

$\therefore$ after 12 seconds, angular velocity $= \dfrac{12 \times 0.3}{0.225} = 16 \text{ rad s}^{-1}$.

(b) K.E. of disc after 12 seconds $= \frac{1}{2}I\omega^2$
$$= \frac{1}{2} \times 0.225 \times 16^2 = 28.8 \text{ J}.$$

(c) Decelerating torque $= 1 \, (\text{N}) \times 0.15 \, (\text{m})$.

$\therefore$ angular deceleration $= \dfrac{\text{torque}}{I} = \dfrac{0.15}{0.225} \text{ rad s}^{-2}$.

$\therefore$ time to bring disc to rest $= \dfrac{\text{initial angular velocity}}{\text{angular deceleration}}$

$$= \dfrac{16 \times 0.225}{0.15} = 24 \text{ seconds}.$$

2. Discuss the importance of the concept of Moment of Inertia in the description of rotational motion.

A uniform Catherine wheel, which may be taken as consisting of many thin circular turns of combustible material, is mounted in a vertical plane so that it may revolve freely about a rail through its centre and perpendicular to its plane. When ignited it produces a steady thrust F and burns its mass away at a constant rate m. If it starts from rest with initial radius a, find its angular velocity when only half its initial mass remains. (*C.S.*)

Let M_0 be the initial mass and suppose the radius has decreased to r from a after a time t. The mass M at this instant $= \dfrac{r^2}{a^2}M_0$.

Couple on wheel $= F \cdot r = I\dfrac{d\omega}{dt}$, where ω is the instantaneous angular velocity.

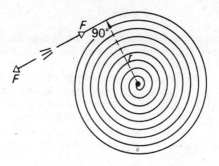

Fig. 3.17 Example

But $I = \dfrac{Mr^2}{2}$.

$$\therefore F = \tfrac{1}{2}Mr\dfrac{d\omega}{dt} \quad . \quad . \quad . \quad . \quad . \quad (1)$$

From previous, $r = (M/M_0)^{1/2} a$. Substituting in (i) for r,

$$\therefore F = \tfrac{1}{2}M\dfrac{d\omega}{dt}\left(\dfrac{M}{M_0}\right)^{1/2} a$$

$$= \tfrac{1}{2}\dfrac{aM^{3/2}}{M_0^{1/2}} \cdot \dfrac{d\omega}{dM} \cdot \dfrac{dM}{dt}.$$

Now $\dfrac{dM}{dt} = -m$, since $\dfrac{dM}{dt}$ is the rate of loss of M.

$$\therefore F = \tfrac{1}{2}\dfrac{aM^{3/2}}{M_0^{1/2}} \cdot \dfrac{d\omega}{dM} \cdot -m.$$

We now re-arrange the variables M and ω, and integrate with respective limits of $M_0/2$, M_0, and ω_1, 0, where ω_1 is the angular velocity after half the mass is burned. Then

$$-\dfrac{2M_0^{1/2}F}{ma}\int_{M_0}^{M_0/2}\dfrac{dM}{M^{3/2}} = \int_0^{\omega_1} d\omega.$$

$$\therefore +\dfrac{2M_0^{1/2}F \times 2}{ma}\left[\left(\dfrac{M_0}{2}\right)^{-1/2} - M_0^{-1/2}\right] = \omega_1.$$

$$\therefore +\dfrac{4M_0^{1/2}F}{ma}\left[\dfrac{\sqrt{2}-1}{M_0^{1/2}}\right] = \omega_1.$$

$$\therefore \omega_1 = \dfrac{4F}{ma}(\sqrt{2}-1).$$

ROTATION OF RIGID BODIES

EXERCISES 3

(*Assume $g = 10$ m s^{-2} unless otherwise stated*)

What are the missing words in the statements 1–4?

1. The kinetic energy of an object rotating about an axis is calculated from ...

2. The angular momentum of the object is calculated from ...

3. '$I \times$ angular acceleration' is equal to the ... on the object.

4. The period of oscillation of an object about an axis is calculated from ...

Which of the following answers, A, B, C, D or E, do you consider is the correct one in the statements 5–8?

5. When a sphere of moment of inertia I about its centre of gravity, and mass m, rolls from rest down an inclined plane without slipping, its kinetic energy is calculated from A $\frac{1}{2}I\omega^2$, B $\frac{1}{2}mv^2$, C $I\omega + mv$, D $\frac{1}{2}I\omega^2 + \frac{1}{2}mv^2$, E $I\omega$.

6. If a hoop of radius a oscillates about an axis through its circumference perpendicular to its plane, the period is A $2\pi\sqrt{a/g}$, B $2\pi\sqrt{2a/g}$, C $2\pi\sqrt{g/a}$, D $2\pi\sqrt{g/2a}$, E $a/2$.

7. Planets moving in orbit round the sun A increase in velocity at points near the sun because their angular momentum is constant, B increase in velocity near the sun because their energy is constant, C decrease in velocity near the sun owing to the increased attraction, D sweep out equal mass in equal times because their energy is constant, E always have circular orbits.

8. If a constant couple of 500 N m turns a wheel of moment of inertia 100 kg m^2 about an axis through its centre, the angular velocity gained in two seconds is A 5 rad s^{-1}, B 100 m s^{-1}, C 200 m s^{-1}, D 2 m s^{-1}, E 10 rad s^{-1}.

9. A uniform rod has a mass of 60 g and a length 20 cm. Calculate the moment of inertia about an axis perpendicular to its length (i) through its centre, (ii) through one end. Prove the formulae used.

10. A horizontal cylinder, moment of inertia 6×10^4 kg m^2 about its axis, rotates steadily with a period of revolution 1·0 s. Find its kinetic energy and angular momentum about the axis.

11. Two constant forces of 2000 N, acting in opposite directions at the end of a diameter 1·5 m long, are suddenly applied tangentially to the cylinder in Qn. 10 to oppose the motion. Find the time taken for the cylinder to come to rest (use angular momentum) and the number of revolutions during this time (use energy).

12. What is the formula for the kinetic energy of (i) a particle, (ii) a rigid body rotating about an axis through its centre of gravity, (iii) a rigid body rotating about an axis through any point? Calculate the kinetic energy of a disc of mass 5 kg and radius 1 m rolling along a plane with a uniform velocity of 2 m s^{-1}.

13. A sphere rolls down a plane inclined at 30° to the horizontal. Find the acceleration and velocity of the sphere after it has moved 5·0 m from rest along

the plane, assuming the moment of inertia of a sphere about a diameter is $2Ma^2/5$, where M is the mass and a is the radius.

14. A uniform rod of length 3·0 m is suspended at one end so that it can move about an axis perpendicular to its length, and is held inclined at 60° to the vertical and then released. Calculate the angular velocity of the rod when (i) it is inclined at 30° to the vertical, (ii) reaches the vertical.

15. Define the *moment of inertia* of a rigid object about an axis.

A ring of radius 2·0 m oscillates about an axis on its circumference which is perpendicular to the plane of the ring. Calculate the period of oscillation. Give an explanation of any formula used.

16. A flywheel with an axle 1·0 cm in diameter is mounted in frictionless bearings and set in motion by applying a steady tension of 2 N to a thin thread wound tightly round the axle. The moment of inertia of the system about its axis of rotation is $5·0 \times 10^{-4}$ kg m². Calculate (*a*) the angular acceleration of the flywheel when 1 m of thread has been pulled off the axle, (*b*) the constant retarding couple which must then be applied to bring the flywheel to rest in one complete turn, the tension in the thread having been completely removed. (*N*.)

17. Define the moment of inertia of a body about a given axis. Describe how the moment of inertia of a flywheel can be determined experimentally.

A horizontal disc rotating freely about a vertical axis makes 100 r.p.m. A small piece of wax of mass 10 g falls vertically on to the disc and adheres to it at a distance of 9 cm from the axis. If the number of revolutions per minute is thereby reduced to 90, calculate the moment of inertia of the disc. (*N*.)

18. Describe an experiment using a bar pendulum to determine the acceleration due to gravity. Show how the result is calculated from the observations.

A uniform disc of diameter 12·0 cm and mass 810 g is suspended with its plane horizontal by a torsion wire and allowed to perform small torsional oscillations about a vertical axis through its centre. The disc is then replaced by a uniform sphere which is allowed to oscillate similarly about a diameter. If the period of oscillation of the sphere is 1·66 times that of the disc, determine the moment of inertia of the sphere about a diameter. (*L*.)

19. Define *angular momentum, moment of inertia*. Explain why, in driving machinery, a flywheel is usually used with a stationary steam engine or gas engine but not with an electric motor.

A constant couple of $2·45 \times 10^{-2}$ N m applied for 30·0 s to a pivoted wheel causes the rate of rotation of the wheel to increase uniformly from zero to 2·50 rev/s. The external couple is then removed and the wheel comes to rest in a further 80·0 s. Find (*a*) the moment of inertia of the wheel about its axis, (*b*) the frictional couple at the bearings, (*c*) the total number of revolutions made by the wheel, (*d*) the greatest kinetic energy of the wheel. (*L*.)

20. Define *moment of inertia* and derive an expression for the kinetic energy of a rigid body of moment of inertia I about a given axis when it is rotating about that axis with a uniform angular velocity ω.

Give two examples of physical phenomena in which moment of inertia is a necessary concept for a theoretical description, in each case showing how the concept is applied.

A uniform spherical ball starts from rest and rolls freely without slipping down an inclined plane at 10° to the horizontal along a line of greatest slope. Calculate its velocity after it has travelled 5 m. (M.I. of sphere about a diameter = $2Mr^2/5$.) (*O. & C.*)

21. Explain the meaning of the term *moment of inertia*. Describe in detail how you would find experimentally the moment of inertia of a bicycle wheel about the central line of its hub.

A uniform cylinder 20 cm long, suspended by a steel wire attached to its mid-point so that its long axis is horizontal, is found to oscillate with a period of 2 seconds when the wire is twisted and released. When a small thin disc, of mass 10 g, is attached to each end the period is found to be 2·3 seconds. Calculate the moment of inertia of the cylinder about the axis of oscillation. (*N.*)

22. What is meant by 'moment of inertia'? Explain the importance of this concept in dealing with problems concerning rotating bodies.

Describe, with practical details, how you would determine whether a given cylindrical body were hollow or not without damaging it. (*C.*)

23. Define *moment of inertia*, and find an expression for the kinetic energy of a rigid body rotating about a fixed axis.

A sphere, starting from rest, rolls (without slipping) down a rough plane inclined to the horizontal at an angle of 30°, and it is found to travel a distance of 13·5 m in the first 3 seconds of its motion. Assuming that F, the frictional resistance to the motion, is independent of the speed, calculate the ratio of F to the *weight* of the sphere. (For a sphere of mass m and radius r, the moment of inertia about a diameter is $\frac{2}{5}mr^2$.) (*O. & C.*)

24. What is meant by angular momentum? Under what conditions is the angular momentum of a system about a given axis conserved?

A small meteorite of mass m travelling towards the centre of the earth strikes the earth at the equator. Assuming that the earth is a uniform sphere of mass M and radius R, show that the length of the day is consequently increased by approximately $5m\tau/(2M)$ seconds where τ is the duration of the day, and that the rotational energy of the system is decreased by approximately $2\pi^2 mR^2/\tau^2$. How do you account for this loss of rotational energy?

What mass of meteorites and meteorite dust falling uniformly on the earth with no net angular momentum would increase the length of the day by one thousandth of a second? (The moment of inertia of a uniform sphere of mass m and radius r about a diameter is $2mr^2/5$. Mass of earth = $6·0 \times 10^{24}$ kg.) (*O. & C.*)

Chapter 4

STATIC BODIES. FLUIDS

STATIC BODIES

Statics

1. STATICS is a subject which concerns the *equilibrium* of forces, such as the forces which act on a bridge. In Fig. 4.1 (i), for example, the joint O of a light bridge is in equilibrium under the action of the two forces P, Q acting in the girders meeting at O and the reaction S of the masonry at O.

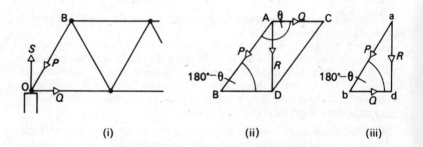

FIG. 4.1 Equilibrium of forces

Parallelogram of Forces

A force is a vector quantity, i.e., it can be represented in magnitude and direction by a straight line (p. 1). If AB, AC represent the forces P, Q respectively at the joint O, their *resultant*, R, is represented in magnitude and direction by the diagonal AD of the parallelogram ABDC which has AB, AC as two of its adjacent sides, Fig. 4.1 (ii). This is known as the *parallelogram of forces*, and is exactly analogous to the parallelogram of velocities discussed on p. 8. Alternatively, a line ab may be drawn to represent the vector P and bd to represent Q, in which case ad represents the resultant R.

By trigonometry for triangle ABD, we have

$$AD^2 = BA^2 + BD^2 - 2BA \cdot BD \cos ABD.$$
$$\therefore R^2 = P^2 + Q^2 + 2PQ \cos \theta,$$

STATIC BODIES, FLUIDS

where θ = angle BAC; the angle between the forces P, Q, = $180°$ − angle ABD. This formula enables R to be calculated when P, Q and the angle between them are known. The angle BAD, or α say, between the resultant R and the force P can then be found from the relation

$$\frac{R}{\sin \theta} = \frac{Q}{\sin \alpha},$$

applying the sine rule to triangle ABD and noting that angle ABD = $180° - \theta$.

Resolved component. On p. 9 we saw that the effective part, or resolved component, of a vector quantity X in a direction θ inclined to it is given by $X \cos \theta$. Thus the resolved component of a force P in a direction making an angle of $30°$ with it is $P \cos 30°$; in a perpendicular direction to the latter the resolved component is $P \cos 60°$, or $P \sin 30°$. In Fig. 4.1 (i), the downward component of the force P on the joint of O is given by $P \cos$ BOS.

Forces in Equilibrium. Triangle of Forces

Since the joint O is in equilibrium, Fig. 4.1 (i), the resultant of the forces P, Q in the rods meeting at this joint is equal and opposite to the reaction S at O. Now the diagonal AD of the parallelogram ABDC in Fig. 4.1 (ii) represents the resultant R of P, Q since ABDC is the parallelogram of forces for P, Q; and hence DA represents the force S. Consequently the sides of the triangle ABD represent the three forces at O in magnitude and direction: This result can be generalised as follows. *If three forces are in equilibrium, they can be represented by the three sides of a triangle taken in order.* This theorem in Statics is known as the *triangle of forces*. In Fig. 4.1 (ii), AB, BD, DA, in this order, represent, P, Q, S respectively in Fig. 4.1 (i)

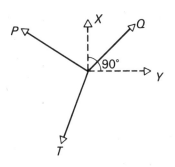

FIG. 4.2 Resolution of forces

We can derive another relation between forces in equilibrium. Suppose X, Y are the respective algebraic sums of the resolved components in two perpendicular directions of three forces P, Q, T in equilibrium, Fig. 4.2. Then, since X, Y can each be represented by the sides of a *rectangle* drawn to scale, their resultant R is given by

$$R^2 = X^2 + Y^2 \quad . \quad . \quad . \quad . \quad \text{(i)}$$

Now if the forces are in equilibrium, R is zero. It then follows from (i) that X must be zero and Y must be zero. Thus *if forces are in equilibrium the algebraic sum of their resolved components in any two perpendicular directions is respectively zero.* This result applies to any number of forces in equilibrium.

EXAMPLE

State what is meant by *scalar* and *vector* quantities, giving examples of each.

Explain how a flat kite can be flown in a wind that is blowing horizontally. The line makes an angle of 30° with the vertical and is under a tension of 0·1 newton; the mass of the kite is 5 g. What angle will the plane of the kite make with the vertical, and what force will the wind exert on it? (*O. & C.*)

Second part. When the kite AB is inclined to the horizontal, the wind blowing horizontally exerts an upward force F normal to AB, Fig. 4.3. For equilibrium of the kite, F must be equal and opposite to the resultant of the tension T, 0·1 newton, and the weight W, 0.005×9.8 or 0·049 N. By drawing the parallelogram of forces for the resultant of T and W, F and the angle θ between F and T can be found. θ is nearly 10°, and F is about 0·148 N. The angle between AB and the vertical $= 60° + \theta = 70°$ (approx.). Also, since F is the component of the horizontal force P of the wind.

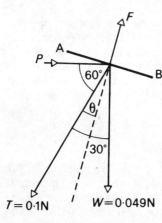

FIG. 4.3 Example

$$P \cos(60° + \theta) = F.$$

$$\therefore P = \frac{F}{\cos(60° + \theta)} = \frac{0.148}{\cos 70°}$$

$$= 0.43 \text{ N}.$$

Moments

When the steering-wheel of a car is turned, the applied force is said to exert a *moment*, or turning-effect, about the axle attached to the wheel. The magnitude of the moment of a force P about a point O is defined as *the product of the force P and the perpendicular distance OA from O to the line of action of P*, Fig. 4.4 (i). Thus

$$\text{moment} = P \times \text{AO}.$$

The magnitude of the moment is expressed in *newton metre* (N m) when P is in newtons and AO is in metres. We shall take an anticlockwise moment as positive in sign and a clockwise moment as negative in sign.

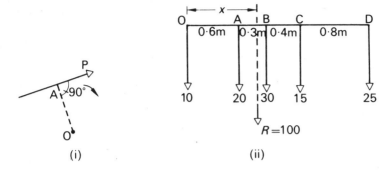

FIG. 4.4 Parallel forces

Parallel Forces

If a rod carries loads of 10, 20, 30, 15, and 25 N at O, A, B, C, D respectively, the resultant R of the weights, which are parallel forces, is equal to their sum, Fig. 4.4 (ii). Thus

$$\text{resultant, } R, = 10+20+30+15+25 = 100 \text{ N}.$$

Experiment and theory show that *the moment of the resultant of a number of forces about any point is equal to the algebraic sum of the moments of the individual forces about the same point*. This result enables us to find where the resultant R acts. Taking moments about O for all the forces in Fig. 4.4 (ii), we have

$$(20 \times 0·6)+(30 \times 0·9)+(15 \times 1·3)+(25 \times 2·1),$$

because the distances between the forces are 0·6 m, 0·3 m, 0·4 m, 0·8 m, as shown. If x m is the distance of the line of action of R from O, the moment of R about O $= R \times x = 100 \times x$.

$$\therefore 100x = (20 \times 0·6)+(30 \times 0·9)+(15 \times 1·3)+(25 \times 2·1),$$

from which $\quad x = 1·1$ m.

Equilibrium of Parallel Forces

The resultant of a number of forces *in equilibrium* is zero; and the moment of the resultant about any point is hence zero. It therefore follows that the algebraic sum of the moments of all the forces about any point is zero when those forces are in equilibrium. This means that the total clockwise moment of the forces about any point = the total anticlockwise moment of the remaining forces about the same point.

As a simple example of the equilibrium of parallel forces, suppose a

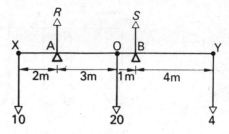

Fig. 4.5 Example

light beam XY rests on supports, A, B, and has loads of 10, 20, and 4 N concentrated at X, O, Y respectively, Fig. 4.5. Let R, S be the reactions at A, B respectively. Then, for equilibrium in a vertical direction,

$$R + S = 10 + 20 + 4 = 34 \text{ N} \qquad . \qquad . \qquad . \qquad \text{(i)}$$

To find R, we take moments about a suitable point such as B, in which case the moment of S is zero. Then, for the remaining four forces,

$$+10.6 + 20.1 - R.4 - 4.4 = 0,$$

from which $R = 16$ N. From (i), it follows that $S = 34 - 16 = 18$ N.

Equilibrium of Three Coplanar Forces

If any object is in equilibrium under the action of *three* forces, the resultant of two of the forces must be equal and opposite to the third force. Thus the line of action of the third force must pass through the point of intersection of the lines of action of the other two forces.

As an example of calculating unknown forces in this case, suppose that a 12 m ladder of 20·0 kgf is placed at an angle of 60° to the horizontal, with one end B leaning against a smooth wall and the other end A on the ground, Fig. 4.6. The force R at B on the ladder is called the *reaction* of the wall, and if the latter is smooth, R acts perpendicularly to the wall. Assuming the weight, W, of the ladder acts at its mid-point G, the forces W and R meet at O, as shown. Consequently the frictional force F at A passes through O.

The *triangle of forces* can be used to find the unknown forces R, F. Since DA is parallel to R, AO is parallel to F, and OD is parallel to W, the triangle of forces is represented by AOD. By means of a scale drawing R and F can be found, since

$$\frac{W(20)}{\text{OD}} = \frac{F}{\text{AO}} = \frac{R}{\text{DA}}.$$

A quicker method is to take moments about A for all the forces. The

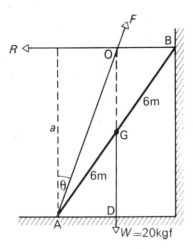

Fig. 4.6 Triangle of forces

algebraic sum of the moments is zero about any point since the object is in equilibrium, and hence

$$R \cdot a - W \cdot AD = 0,$$

where a is the perpendicular from A to R. (F has zero moment about A.) But $a = 12 \sin 60°$, and $AD = 6 \cos 60°$.

$$\therefore R \times 12 \sin 60° - 20 \times 6 \cos 60° = 0.$$

$$\therefore R = 10 \frac{\cos 60°}{\sin 60°} = 5 \cdot 8 \text{ kgf.}$$

Suppose θ is the angle F makes with the vertical.

Resolving the forces vertically, $F \cos \theta = W = 20$ kgf.

Resolving horizontally, $F \sin \theta = R = 5 \cdot 8$ kgf.

$$\therefore F^2 \cos^2 \theta + F^2 \sin^2 \theta = F^2 = 20^2 + 5 \cdot 8^2.$$

$$\therefore F = \sqrt{20^2 + 5 \cdot 8^2} = 20 \cdot 8 \text{ kgf.}$$

Couples and Torque

There are many examples in practice where two forces, acting together, exert a moment or turning-effect on some object. As a very simple case, suppose two strings are tied to a wheel at X, Y, and *two equal and opposite forces*, F, are exerted tangentially to the wheel, Fig. 4.7 (i). If the wheel is pivoted at its centre, O, it begins to rotate about O in an anticlockwise direction.

Two equal and opposite forces whose lines of action do not coincide are said to constitute a *couple* in Mechanics. The two forces always have a turning-effect, or moment, called a *torque*, which is defined by

$$\text{torque} = \text{one force} \times \text{perpendicular distance between forces} \quad (1)$$

Since XY is perpendicular to each of the forces F in Fig. 46 (i), the

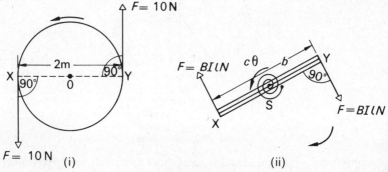

FIG. 4.7 Couple and torque

moment of the couple acting on the wheel $= F \times XY = F \times$ diameter of wheel. Thus if $F = 10$ newton and the diameter is 2 metre, the moment of the couple or torque $= 20$ newton metre (N m).

In the theory of the *moving-coil electrical instrument*, we meet a case where a coil rotates when a current I is passed into it and comes to rest after deflection through an angle θ. Fig. 4.7 (ii). The forces F on the two sides X and Y of the coil are both equal to $BIlN$, where B is the strength of the magnetic field, l is the length of the coil and N is the number of turns (see Electricity (*Arnold*) by the author). Thus the coil is deflected by a couple. The moment or torque of the deflecting couple $= F \times b$, where $b = XY =$ breadth of coil. Hence

$$\text{torque} = BIlN \times b = BANI,$$

where $A = lb =$ area of coil. The opposing couple, due to the spring S, is $c\theta$, where c is its elastic constant (p. 192). Thus, for equilibrium, $BANI = c\theta$.

Work Done by a Couple

Suppose two equal and opposite forces F act tangentially to a wheel W, and rotate it through an angle θ while the forces keep tangentially to the wheel, Fig. 4.8. The moment of the couple is then constant.

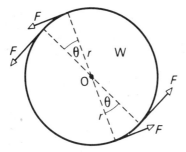

Fig. 4.8 Work done by couple

The work done by each force = $F \times$ distance = $F \times r\theta$, since $r\theta$ is the distance moved by a point on the rim if θ is in radians.

∴ total work done by couple = $Fr\theta + Fr\theta = 2Fr\theta$.

But moment of couple = $F \times 2r = 2Fr$

∴ *work done by couple* = *torque* or *moment of couple* $\times \theta$

Although we have chosen a simple case, the result for the work done by a couple is always given by *torque × angle of rotation*. In the formula, it should be carefully noted that θ is in radians. Thus suppose $F = 20$ newton, $r = 4$ cm $= 0.04$ metre, and the wheel makes 5 revolutions while the moment of the couple is kept constant. Then, from above,

torque or moment of couple = 20×0.08 newton metre.

and angle of rotation = $2\pi \times 5$ radian.

∴ work done = $20 \times 0.08 \times 2\pi \times 5 = 50$ J

Centre of Gravity

Every particle is attracted towards the centre of the earth by the force of gravity, and the *centre of gravity* of a body is the point where the *resultant* force of attraction or *weight* of the body acts. In the simple case of a ruler, the centre of gravity is the point of support when the ruler is balanced. A similar method can be used to find roughly the centre of gravity of a flat plate. A more accurate method consists of suspending the object in turn from two points on it, so that it hangs freely in each case, and finding the point of intersection of a plumb-line, suspended in turn from each point of suspension. This experiment is described in elementary books.

An object can be considered to consist of many small particles. The forces on the particles due to the attraction of the earth are all parallel since they act vertically, and hence their resultant is the sum of all the forces. The resultant is the *weight* of the whole object, of course. In the case of a rod of uniform cross-sectional area, the weight of a particle A at one end, and that of a corresponding particle A' at the other end, have a resultant which acts at the mid-point O of the rod, Fig. 4.9 (i).

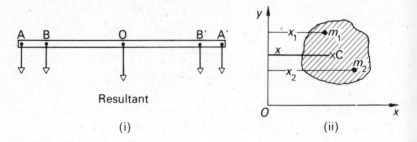

FIG. 4.9   Centre of gravity and mass

Similarly, the resultant of the weight of a particle B, and that of a corresponding particle at B', have a resultant acting at O. In this way, i.e., by symmetry, it follows that the resultant of the weights of all the particles of the rod acts at O. Hence the centre of gravity of a uniform rod is at its mid-point.

The centre of gravity, C.G., of the curved surface of a hollow cylinder acts at the midpoint of the cylinder axis. This is also the position of the C.G. of a uniform solid cylinder. The C.G. of a triangular plate or lamina is two-thirds of the distance along a median from corresponding point of the triangle. The C.G. of a uniform right solid cone is three-quarters along the axis from the apex.

Centre of Mass

The 'centre of mass' of an object is the point where its total mass acts or appears to act. Fig. 4.9 (ii) illustrates how the position of the centre of mass of an object may be calculated, using axes Ox, Oy.

If m_1 is the mass of a small part of the object and x_1 is the perpendicular distance to Oy, then $m_1 x_1$ represents a product similar to the moment of a weight at m_1 about Oy. Likewise, $m_2 x_2$ is a 'moment' about Oy, where m_2 is another small part of the object. The sum of the total 'moments' about Oy of all the parts of the object can be written Σmx. The total mass $= \Sigma m = M$ say. The distance $\bar{x}$ of the centre of mass C from Oy is then given by

$$\bar{x} = \frac{\Sigma mx}{M}.$$

Similarly, the distance $\bar{y}$ of the centre of mass C from Ox is given by

$$\bar{y} = \frac{\Sigma my}{M}.$$

If the earth's field is uniform at all parts of the body, then the *weight* of a small mass m of it is typically mg. Thus, by moments, the distance of the centre of gravity from Oy is given by

$$\frac{\Sigma mg \times x}{\Sigma mg} = \frac{\Sigma mx}{\Sigma m} = \frac{\Sigma mx}{M}.$$

The acceleration due to gravity, g, cancels in numerator and denominator. It therefore follows that the centre of mass *coincides* with the centre of gravity. However, if the earth's field is *not* uniform at all parts of the object, the weight of a small mass m_1 is then $m_1 g_1$ say and the weight of another small mass m_2 is $m_2 g_2$. Clearly, the centre of gravity does not now coincide with the centre of mass. A very long or large object has different values of g at various parts of it.

EXAMPLE

What is meant by (a) the centre of mass of a body, (b) the centre of gravity of a body?

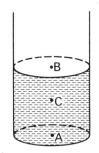

FIG. 4.10 Example

A cylindrical can is made of a material of mass 10 g cm^{-2} and has no lid. The diameter of the can is 25 cm and its height 50 cm. Find the position of the centre of mass when the can is half full of water. (*C*.)

The area of the base $= \pi r^2 = \pi \times (25/2)^2$ cm^2; hence the mass is $\pi \times (25/2)^2 \times 10$ g, and acts at A, the centre of the base, Fig. 4.10.

The mass of the curved surface of the centre $= 2\pi rh \times 10$ g $= 2\pi \times (25/2) \times 50 \times 10$ g, and acts at B, half-way along the axis.

The mass of water $= \pi r^2 h$ g $= \pi \times (25/2)^2 \times 25$ g, and acts at C, the mid-point of AB.

Thus the resultant mass in gramme

$$= \frac{\pi \times 625 \times 10}{4} + \frac{2\pi \times 25 \times 50 \times 10}{2} + \frac{\pi \times 625 \times 25}{4}$$

$$= \pi \times 625 \times 28\tfrac{3}{4}.$$

Taking moments about A,

$$\therefore \pi \times 625 \times 28\tfrac{3}{4} \times x = (\pi \times 12500) \times \text{AB} + \left(\pi \times \frac{625 \times 25}{4}\right) \times \text{AC}$$

where x is the distance of the centre of mass from A.

$$\therefore 28\tfrac{3}{4}x = 20 \times 25 + \frac{25}{4} \times 12\tfrac{1}{2}$$

$$\therefore x = 20 \text{ (approx)}.$$

$\therefore$ centre of mass is 20 cm from the base.

Types of Equilibrium

If a marble A is placed on the curved surface of a bowl S, it rolls

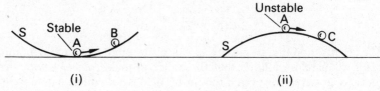

Fig. 4.11 Stable and unstable equilibrium

down and settles in equilibrium at the lowest point. Fig. 4.11 (i). *Its potential energy is then a minimum.* This is the case for objects in any field, gravitational, magnetic or electrical. The equilibrium position corresponds to minimum potential energy.

If the marble A is disturbed and displaced to B, its energy increases. When it is released, the marble rolls back to A. Thus the marble at A is said to be in *stable equilibrium*. Note that the centre of gravity of A is *raised* on displacement to B. On this account the forces in the field return the marble from B to A, where its potential energy is lower.

Suppose now that the bowl S is inverted and the marble is placed at its top point at A. Fig. 4.11 (ii). If A is displaced slightly to C, its potential energy and centre of gravity are then *lowered*. A now continues to move further away from B under the action of the forces in the field. Thus in Fig. 4.10 (ii), A is said to be in *unstable equilibrium*.

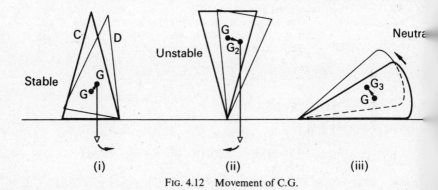

Fig. 4.12 Movement of C.G.

STATIC BODIES, FLUIDS

Fig. 4.12 (i) shows a cone C with its base on a horizontal surface. If it is slightly displaced to D, its centre of gravity G rises to G_1. As previously explained, D returns to C when the cone is released, so that the equilibrium is stable. In Fig. 4.12 (ii), the cone is balanced on its apex. When it is slightly displaced, the centre of gravity, G_1 is *lowered* to G_2. This is unstable equilibrium. Fig. 4.12 (iii) illustrates the case of the cone resting on its curved surface. If it is slightly displaced, the centre of gravity G remains at the same height G_3. The cone hence remains in its displaced position. This is called *neutral equilibrium*.

EXAMPLE

A rectangular beam of thickness a is balanced on the curved surface of a rough cylinder of radius r. Show that the beam is stable if r is greater than $a/2$.

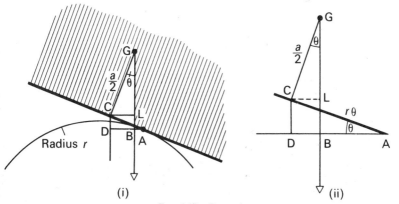

FIG. 4.13 Example

Suppose the beam is tilted through a small angle θ. The point of contact C then moves to A, the radius of the cylinder moves through an angle θ, and the vertical GB through the centre of gravity G of the beam makes an angle θ with CG. (Fig. 4.13 (i). As shown in the exaggerated sketch in Fig. 4.13 (ii), AC = $r\theta$.

The beam is in stable equilibrium if the vertical through G lies to the left of A, since a restoring moment is then exerted. Thus for stable equilibrium, AD must be greater than DB, where CD is the vertical through C.

Now
$$AD = r\theta \cos \theta, \quad DB = CL = \frac{a}{2} \sin \theta.$$
$$\therefore r\theta \cos \theta > \frac{a}{2} \sin \theta.$$

When θ is very small, $\cos \theta \to 1$, $\sin \theta \to \theta$.
$$\therefore r\theta > \frac{a}{2} \theta.$$
$$\therefore r > \frac{a}{2}.$$

Common Balance

The common balance is basically a lever whose two arms are equal, Fig. 4.14. The fulcrum, about which the beam and pointer tilt, is an agate wedge resting on an agate plate; agate wedges, B, at the ends of the beam, support the scale-pans. The centre of gravity of the beam and

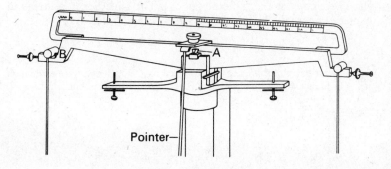

Fig. 4.14 Common balance

pointer is vertically below the fulcrum, to make the arrangement stable. The masses placed on the two scale-pans are equal when there is a 'balance'.

On rare occasions the arms of the balance are slightly unequal. The mass W of an object is then determined by finding the respective masses W_1, W_2 required to balance it on each scale-pan. Suppose a, b are the lengths of the respective arms. Then, taking moments,

$$\therefore W_1 \cdot a = W \cdot b, \text{ and } W \cdot a = W_2 \cdot b.$$

$$\therefore \frac{W}{W_1} = \frac{a}{b} = \frac{W_2}{W}$$

$$\therefore W^2 = W_1 W_2$$

$$\therefore W = \sqrt{W_1 W_2}.$$

Thus W can be found from the two masses W_1, W_2.

Sensitivity of a Balance

A balance is said to be very *sensitive* if a small difference in weights on the scale-pans causes a large deflection of the beam. To investigate the factors which affect the sensitivity of a balance, suppose a weight W_1 is placed on the left scale-pan and a slightly smaller weight W_2 is

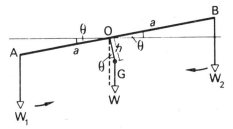

Fig. 4.15 Theory of balance

placed on the right scale-pan, Fig. 4.15. The beam AOB will then be inclined at some angle θ to the horizontal, where O is the fulcrum.

The weight W of the beam and pointer acts at G, at a distance h below O. Suppose $AO = OB = a$. Then, taking moments about O,

$$W_1 a \cos \theta = Wh \sin \theta + W_2 a \cos \theta$$
$$\therefore (W_1 - W_2) a \cos \theta = Wh \sin \theta$$
$$\therefore \tan \theta = \frac{(W_1 - W_2)a}{Wh}.$$

Thus for a given value of $(W_1 - W_2)$, the difference of the weights on the scale-pans, θ will increase when a increases and W, h both decrease. In theory, then, a sensitive balance must be light and have long arms, and the centre of gravity of its beam and pointer must be very close to the fulcrum. Now a light beam will not be rigid. Further, a beam with long arms will take a long time to settle down when it is deflected. A compromise must therefore be made between the requirements of sensitivity and those of design.

If the knife-edges of the scale-pan and beam are in the same plane, corresponding to A, B and O in Fig. 4.15, then the weights W_1, W_2 on them always have the same perpendicular distance from O, irrespective of the inclination of the beam. In this case the net moment about O is $(W_1 - W_2)a \cos \theta$. Thus the moment depends on the difference, $W_1 - W_2$, of the weights and not on their actual values. Hence the sensitivity is independent of the actual load value over a considerable range.

When the knife-edge of the beam is below the knife-edges of the two scale-pans, the sensitivity increases with the load; the reverse is the case if the knife-edge of the beam is above those of the scale-pans.

Buoyancy Correction in Weighing

In very accurate weighing, a correction must be made for the buoyancy of the air. Suppose the body weighed has a density ρ and a mass m. From Archimedes principle (p. 132), the upthrust due to the air of density σ is equal to the weight of air displaced by the body, and hence the net downward force $= \left(m - \dfrac{m}{\rho} \cdot \sigma\right) g$, since the volume of the body is m/ρ. Similarly, if the weights restoring a balance have a total mass m_1 and a density ρ_1, the net downward force $= \left(m_1 - \dfrac{m_1}{\rho_1} \cdot \sigma\right) g$. Since there is a balance,

$$m - \frac{m\sigma}{\rho} = m_1 - \frac{m_1 \sigma}{\rho_1}$$

$$\therefore m = m_1 \frac{\left(1 - \dfrac{\sigma}{\rho_1}\right)}{1 - \dfrac{\sigma}{\rho}}.$$

Thus knowing the density of air, σ, and the densities ρ, ρ_1, the true mass m can be found in terms of m_1. The pressure and temperature of air, which may vary from day to day, affects the magnitude of its density σ, from the gas laws; the humidity of the air is also taken into account in very accurate weighing, as the density of moist air differs from that of dry air.

FLUIDS

Pressure

Liquids and gases are called *fluids*. Unlike solid objects, fluids can flow.

If a piece of cork is pushed below the surface of a pool of water and then released, the cork rises to the surface again. The liquid thus exerts an upward force on the cork and this is due to the *pressure* exerted on the cork by the surrounding liquid. Gases also exert pressures. For example, when a thin closed metal can is evacuated, it

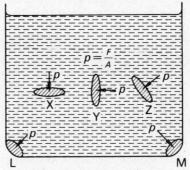

Fig. 4.16 Pressure in liquid

usually collapses with a loud explosion. The surrounding air now exerts a pressure on the outside which is no longer counter-balanced by the pressure inside, and hence there is a resultant force.

Pressure is defined as the *average force per unit area* at the particular region of liquid or gas. In Fig. 4.16, for example, X represents a small horizontal area, Y a small vertical area and Z a small inclined area, all inside a vessel containing a liquid. The pressure p acts normally to the planes of X, Y or Z. In each case

$$\text{average pressure, } p, = \frac{F}{A},$$

where F is the normal force due to the liquid on one side of an area A of X, Y or Z. Similarly, the pressure p on the sides L or M of the curved vessel acts normally to L and M and has magnitude F/A. In the limit, when the area is very small, $p = dF/dA$.

At a given point in a liquid, the pressure can act in any direction. Thus *pressure is a scalar*, not a vector. The direction of the force on a particular surface is normal to the surface.

Formula for Pressure

Observation shows that the pressure increases with the depth, h, below the liquid surface and with its density ρ.

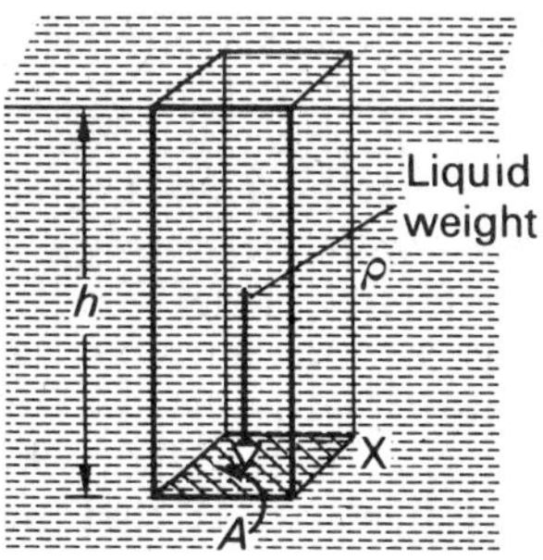

To obtain a formula for the pressure, p, suppose that a horizontal plate X of area A is placed at a depth h below the liquid surface, Fig. 4.17. By drawing vertical lines from points on the perimeter of X, we can see that the force on X due to the liquid is equal to the weight of liquid of height h and uniform cross-section A. Since the volume of this liquid is Ah, the mass of the liquid $= Ah \times \rho$.

FIG. 4.17 Pressure and depth

$$\therefore \text{weight} = Ah\rho g \text{ newton},$$

where g is 9·8, h is in m, A is in m², and ρ is in kg m⁻³.

$$\therefore \text{pressure, } p, \text{ on } X = \frac{\text{force}}{\text{area}} = \frac{Ah\rho g}{A}$$

$$\therefore \boxed{p = h\rho g} \quad . \quad . \quad . \quad . \quad (1)$$

When h, ρ, g have the units already mentioned, the pressure p is in *newton metre⁻²* (N m⁻²).

The *bar* is a unit of pressure used in meteorology. By definition,

$$1 \text{ bar} = 10^5 \text{ N m}^{-2} \quad . \quad . \quad . \quad (2)$$

The *pascal* (Pa) is the name given to a pressure of 1 N m^{-2}. Thus

$$1 \text{ bar} = 10^5 \text{ Pa} \quad . \quad . \quad . \quad (3)$$

Pressure is often expressed in terms of that due to a height of mercury (Hg). One unit is the *torr* (after Torricelli):

$$1 \text{ torr} = 1 \text{ mmHg} = 133 \cdot 3 \text{ N m}^{-2} \text{ (approx)}.$$

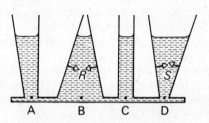

Fig. 4.18 Pressure and cross-section

From $p = h\rho g$ it follows that *the pressure in a liquid is the same at all points on the same horizontal level in it.* Experiment also gives the same result. Thus a liquid filling the vessel shown in Fig. 4.18 rises to the same height in each section if ABCD is horizontal. The cross-sectional area of B is greater than that of D; but the force on B is the sum of the weight of water above it together with the downward component of reaction R of the sides of the vessel, whereas the force on D is the weight of water above it *minus* the upward component of the reaction S of the sides of the vessel. It will thus be noted that the pressure in a vessel is independent of the cross-sectional area of the vessel.

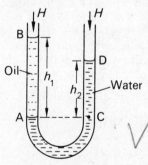

Fig. 4.19
Comparison of densities

Liquids in U-tube

Suppose a U-tube is partly filled with water, and oil is then poured into the left side of the tube. The oil will then reach some level B at a height h_1 above the surface of separation, A, of the water and oil, while the water on the right side of the tube will then reach some level D at a height h_2 above the level of A, Fig. 4.19.

STATIC BODIES, FLUIDS

Since the pressure in the water at A is equal to the pressure at C on the same horizontal level, it follows that

$$H + h_1 \rho_1 g = H + h_2 \rho_2 g,$$

where H is the atmospheric pressure, and ρ_1, ρ_2 are the respective densities of oil and water. Simplifying,

$$h_1 \rho_1 = h_2 \rho_2$$

$$\therefore \rho_1 = \rho_2 \times \frac{h_2}{h_1}.$$

Since ρ_2 (water) $= 1000 \text{ kg m}^{-3}$, and h_2, h_1 can be measured, the density ρ_1 of the oil can be found.

Atmospheric Pressure

The pressure of the atmosphere was first measured by Galileo, who observed the height of a water column in a tube placed in a deep well. About 1640 TORRICELLI thought of the idea of using mercury instead of water, to obtain a much shorter column. He completely filled a glass tube about a metre long with mercury, and then inverted it in a vessel D containing the liquid, taking care that no air entered the tube. He observed that the mercury in the tube fell to a level A about 76 cm or 0·76 m above the level of the mercury in D, Fig. 4.20. Since there was

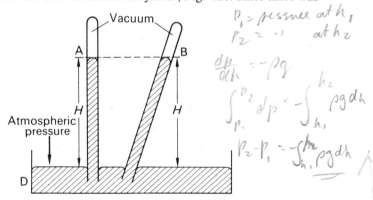

FIG. 4.20 Atmospheric pressure

no air originally in the tube, there must be a vacuum above the mercury at A, neglecting the vapour pressure of the mercury, and it is called a *Torricellian vacuum*.

If the tube in Fig. 4.20 is inclined to the vertical, the mercury ascends the tube to a level B at the same vertical height H above the level of the mercury in D as A.

The pressure on the surface of the mercury in D is atmospheric pressure; and since the pressure is transmitted through the liquid, the atmospheric pressure supports the column of mercury in the tube. Suppose A is at a height H above the level of the mercury in D. Now the pressure, p, at the bottom of a column of liquid of height H and density ρ is given by $p = H\rho g$ (p. 125). Thus if $H = 760$ mm $= 0.76$ m and $\rho = 13600$ kg m^{-3},

$$p = H\rho g = 0.76 \times 13600 \times 9.8 = 1.013 \times 10^5 \text{ N m}^{-2}.$$

The pressure at the bottom of a column of mercury 760 mm high for a particular mercury density and value of g is known as *standard pressure* or *one atmosphere*. By definition, 1 atmosphere $= 1.01325 \times 10^5$ N m^{-2}. *Standard temperature and press* (s.t.p.) is 0°C and 760 mm Hg pressure.

The *bar* is 10^5 N m^{-2}, and is thus very nearly equal to one atmosphere.

Fortin's Barometer

A *barometer* is an instrument for measuring the pressure of the atmosphere, which is required in weather-forecasting, for example. The most accurate form of barometer is due to FORTIN, and like the simple arrangement already described, it consists basically of a barometer tube containing mercury, with a vacuum at the top, Fig. 4.21. One end of the tube dips into a pool of mercury contained in a washleather bag B. A brass scale C graduated in centimetres and millimetres is fixed at the top of the barometer. The zero of the scale correspondings to the tip of an ivory tooth P, and hence, before the level of the top of the mercury is read from the scales, the screw S is adjusted until the level of the mercury in B just reaches the tip of P. A vernier scale V can be moved by a screw D until the bottom of it just reaches the top of the mercury in the tube, and the reading of the height of the mercury is taken from C and V. Torricelli was the first person to observe the variation of the barometric height as the weather changed.

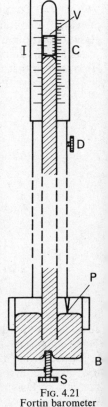

FIG. 4.21
Fortin barometer

'Correction' to the Barometric Height

For comparison purposes, the pressure read on a barometer is often 'reduced' or 'corrected' to the magnitude the pressure would have at 0°C and at sea-level, latitude 45°. Suppose the 'reduced' pressure is H_o cm of mercury, and the observed pressure

STATIC BODIES, FLUIDS

is H_t cm of mercury, corresponding to a temperature of $t°C$. Then, since pressure = $h\rho g$ (p. 125),

$$H_o \rho_o g = H_t \rho_t g',$$

where g is the acceleration due to gravity at sea-level, latitude 45°, and g' is the acceleration at the latitude of the place where the barometer was read.

$$\therefore H_o = H_t \times \frac{\rho_t}{\rho_o} \times \frac{g'}{g}.$$

The magnitude of g'/g can be obtained from standard tables. The ratio ρ_t/ρ_o of the densities = $1/(1+\gamma t)$. where γ is the absolute or true cubic expansivity of mercury. Further, the observed height H_t, on the brass scale requires correction for the expansion of brass from the temperature at which it was correctly calibrated. If the latter is 0°C, then the corrected height is $H_t(1+\alpha t)$, where α is the mean linear expansivity of brass. Thus, finally, the 'corrected' height H_o is given by

$$H_o = H_t \cdot \frac{1+\alpha t}{1+\gamma t} \cdot \frac{g'}{g}.$$

For further accuracy, a correction must be made for the surface tension of mercury (p. 165).

Variation of atmospheric pressure with height

The density of a liquid varies very slightly with pressure. The density of a gas, however, varies appreciably with pressure. Thus at sea-level the density of the atmosphere is about $1·2$ kg m^{-3}; at 1000 m above sea-level the density is about $1·1$ kg m^{-3}; and at 5000 m above sea-level it is about $0·7$ kg m^{-3}. Normal atmospheric pressure is the pressure at the base of a column of mercury 760 mm high, a liquid which has a density of about 13600 kg m^{-3}. Suppose air has a constant density of about $1·2$ kg m^{-3}. Then the height of an air column of this density which has a pressure equal to normal atmospheric pressure

$$= \frac{760}{1000} \times \frac{13600}{1·2} \text{ m} = 8·4 \text{ km}.$$

In fact, the air 'thins' the higher one goes, as explained above. The height of the air is thus much greater than 8·4 km.

Constant temperature atmosphere

Suppose the temperature of the atmosphere is constant. To take account of the variation of density with height, consider a very small height δh of air.

Suppose the air has a density ρ and a pressure p at a height h above the earth's surface. Then, over a further small height δh, the change δp in pressure is given by $\delta p = -\delta h.\rho.g$, the minus indicating that the pressure decreases as the height increases.

$$\therefore \frac{dp}{dh} = -g\rho.$$

Now for a perfect gas of mass m and gas-constant per gram R, $pV = mRT$, with the usual notation, or $p = \rho RT$. Thus $\rho = p/RT$. Substituting for ρ, and assuming the temperature of the atmosphere is constant (isothermal atmosphere),

$$\therefore \frac{dp}{dh} = -\frac{gp}{RT}$$

$$\therefore \int_{p_0}^{p} \frac{dp}{p} = -\frac{g}{RT} \int_{0}^{h} dh,$$

where p_0 is the pressure at the earth's surface, at $h = 0$.

$$\therefore \log_e \left(\frac{p}{p_0}\right) = -\frac{gh}{RT}$$

$$\therefore p = p_0 e^{-gh/RT} \qquad \qquad \qquad \text{(i)}$$

Thus the pressure diminishes exponentially with height. In practice, the atmosphere has constant temperature above 11000 m (*stratosphere*). From sea level to 11000 m, the temperature decreases (*troposphere*).

EXAMPLE

Assuming the temperature of the atmosphere to be constant, calculate the height at which the barometer stands at 75 cm, if its reading at sea-level is 76 cm. Density of air = 1.3 kg m^{-3}, density of mercury = 13600 kg m^{-3} (C.S.)

Since the density of air = 1.3 kg m^{-3} at S.T.P., 1 kg of air occupies a volume of $1/1.3$ m^3. Hence, from $pV = RT$,

$$RT = pV = \frac{0.76 \times 13600 \times 9.8}{1.3}$$

$$\therefore \frac{RT}{g} = \frac{0.76 \times 13600}{1.3} \qquad \qquad \qquad \text{(i)}$$

From our previous formula, $p = p_0 e^{-gh/RT}$,

$$\frac{gh}{RT} = \log_e \left(\frac{p_0}{p}\right),$$

or $$h = \frac{RT}{g} \log_e \left(\frac{p_0}{p}\right).$$

Since $(p_0/p) = (76/75)$, it follows with (i) that

$$h = \frac{0.76 \times 13600}{1.3} \times \log_e \left(\frac{76}{75}\right)$$

$$= 104.6 \text{ m}.$$

STATIC BODIES, FLUIDS   131

Brownian motion. Perrin's experiment

In 1827 Brown, a botanist, observed through a microscope that pollen particles in suspension were moving about constantly in an irregular manner. It became known later that this was due to the ceaseless bombardment of the particles by the molecules of the liquid. The resultant force is unbalanced and random taken over a time, and the phenomenon is known as *Brownian movement* or *motion*.

Perrin, a French scientist, performed a series of brilliant researches on Brownian motion in 1910. He considered that the particles in suspension were moving about inside the liquid in a similar way to molecules. Consequently the particles could be considered as a 'gas'. A column of gas such as the atmosphere settles with a different number of molecules per unit volume or pressure at different heights, as we have just seen. Perrin considered that the particles would also settle in a liquid with different concentrations at different levels.

Theory of Perrin's Experiment. Consider a number of similar particles of density ρ suspended in a liquid of density σ. If the volume of a particle is v its weight is $v\rho g$ and the upthrust on it is $v\sigma g$. In a volume of height dh and unit area of cross-section, the number of particles is $n \cdot dh$, where n is the number per unit volume. The pressure dp due to these particles is thus given by

$$dp = -v(\rho-\sigma)gn \cdot dh \qquad \qquad \text{(i)}$$

where h is measured positively vertically upwards.

Assuming the pressure p obeys the gas laws, then generally $pV = KT$, where K is the gas constant for the particles. If we consider a unit volume, then $V = 1$. And if n is the number of particles per unit volume and k is the gas constant per molecule, then $K = nk = nR/N$, where N is the number of molecules in a mole and R is the molar gas constant.

Thus
$$p = nkT = n\frac{RT}{N} \qquad \qquad \text{(ii)}$$

$$\therefore dp = dn\frac{RT}{N}$$

From (i), it follows that

$$dn\frac{RT}{N} = -v(\rho-\sigma)gn \cdot dh.$$

$$\therefore \int_{n_1}^{n_2} \frac{dn}{n} = -\frac{Nvg}{RT}(\rho-\sigma)\int_{h_1}^{h_2} dh.$$

$$\therefore \log_e\left(\frac{n_2}{n_1}\right) = \frac{Nvg}{RT}(\rho-\sigma)(h_1-h_2) \qquad \qquad \text{(iii)}$$

where n_1, n_2 are the respective number of particles per unit volume at heights h_1, h_2.

Perrin's Experiment. Perrin used gamboge particles of uniform size of the order of 10^{-3} mm diameter; these were obtained by a process of centrifuging, which took several months of patient work. The emulsion was placed on a microscope slide with a cover glass to form a vertical column about 0·1 mm deep, and observed through a powerful microscope mounted vertically. When the suspension had settled down, a count was made continually of the number n_1 of the particles visible at a depth corresponding to h_1. The microscope was then raised slightly, and a new number of particles n_2 was observed corresponding to a depth h_2. The shift of the microscope was $(h_1 - h_2)$. To find the volume v of a particle, the terminal velocity v_0 of the particles was measured as they were dropped through the liquid, and from Stokes' law, $4\pi a^3(\rho - \sigma)g/3 = 6\pi \eta a v_0$. The radius a of the particle can thus be found, and the volume v determined from $4\pi a^3/3$.

On substituting his results in equation (iii), Perrin found N to be of the order $6 \cdot 8 \times 10^{23}$ per mol. This is in good agreement with the value of N found from the kinetic theory of gases, which is $6 \cdot 0 \times 10^{23}$ per mol, and is very striking evidence in support of the kinetic theory applied to liquids. Before Perrin's experiments many scientists of the day had doubted the existence of molecules and hence the validity of the kinetic theory.

Density. Relative Density

As we have seen, the pressure in a fluid depends on the density of the fluid.

The *density* of a substance is defined as its *mass per unit volume*. Thus

$$\text{density}, \rho, = \frac{\text{mass of substance}}{\text{volume of substance}} \qquad . \qquad (47)$$

The density of copper is about 9·0 g cm^{-3} or 9×10^3 kg m^{-3}; the density of aluminium is 2·7 g cm^{-3} or $2 \cdot 7 \times 10^3$ kg m^{-3}; the density of water at 4°C is 1 g cm^{-3} or 1000 kg m^{-3}.

Substances which float on water have a density less than 1000 kg m^{-3} (p. 135). For example, ice has a density of about 900 kg m^{-3}; cork has a density of about 250 kg m^{-3}. Steel, of density 8500 kg m^{-3}, will float on mercury, whose density is about 13600 kg m^{-3} at 0°C.

The density of a substance is often expressed relative to the density of water. This is called the *relative density* or *specific gravity* of the substance. It is a ratio or number, and has no units. The relative density of mercury is 13·6. Thus the density of mercury is 13·6 times the density of water, 1000 kg m^{-3}, and is hence 13600 kg m^{-3}. Copper has a relative density of 9·0 and hence a density of 9000 kg m^{-3}.

Archimedes' Principle

An object immersed in a fluid experiences a resultant upward force owing to the pressure of fluid on it. This upward force is called the *upthrust* of the fluid on the object. ARCHIMEDES stated that *the upthrust is equal to the weight of fluid displaced by the object*, and this is known as

Archimedes' Principle. Thus if an iron cube of volume 400 cm^3 is totally immersed in water of density 1 g cm^{-3}, the upthrust on the cube = $400 \times 1 = 400$ gf. If the same cube is totally immersed in oil of density 0·8 g cm^{-3}, the upthrust on it = $400 \times 0.8 = 320$ gf.

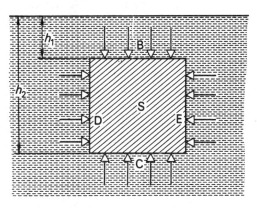

FIG. 4.22 Archimedes' Principle

Fig. 4.22 shows why Archimedes' Principle is true. If S is a solid immersed in a liquid, the pressure on the lower surface C is greater than on the upper surface B, since the pressure at the greater depth h_2 is more than that at h_1. The pressure on the remaining surfaces D and E act as shown. The *force* on each of the four surfaces is calculated by summing the values of *pressure × area* over every part, remembering that vector addition is needed to sum forces. With a simple *rectangular-shaped solid* and the sides, D, E vertical, it can be seen that (i) the resultant horizontal force is zero, (ii) the upward force on C = pressure × area $A = h_2 \rho g A$, where ρ is the liquid density and the downward force on B = pressure × area $A = h_1 \rho g A$. Thus

resultant force on solid = upward force (upthrust) = $(h_2 - h_1)\rho g A$.

But $(h_2 - h_1)A$ = volume of solid, V,

∴ *upthrust* = $V\rho g = mg$, where $m = V\rho$.

∴ *upthrust* = *weight of liquid displaced*.

With a solid of irregular shape, taking into account horizontal and vertical components of forces, the same result is obtained. The upthrust is the weight of *liquid* displaced whatever the nature of the object immersed, or whether it is hollow or not. This is due primarily to the fact that the pressure on the object depends on the liquid in which it is placed.

Density or Relative Density measurement by Archimedes' Principle

The upthrust on an object immersed in water, for example, is the difference between (i) its weight in air when attached to a spring-balance and (ii) the reduced reading on the spring-balance or 'weight' when it is totally immersed in the liquid. Suppose the upthrust is found to be 100 gf. Then, from Archimedes' Principle, the object displaces 100 gf of water. But the density of water is 1 g cm^{-3}. Hence the volume of the object = 100 cm^3, which is numerically equal to the difference in weighings in (i) and (ii).

The density or relative density of a *solid* such as brass or iron can thus be determined by (1) weighing it in air, m_0 gf say, (2) weighing it when it is totally immersed in water, m_1 gf say. Then

$$\text{upthrust} = m_0 - m_1 = \text{wt. of water displaced.}$$

$$\therefore \text{relative density of solid} = \frac{m_0}{m_0 - m_1},$$

and density of solid, $\rho = \frac{m_0}{m_0 - m_1} \times \text{density of water.}$

The density or relative density of a *liquid* can be found by weighing a solid in air (m_0), then weighing it totally immersed in the liquid (m_1), and finally weighing it totally immersed in water (m_2).

Now $m_0 - m_2$ = upthrust in water = weight of water displaced,

and $m_0 - m_1$ = upthrust in liquid = weight of liquid displaced.

$$\therefore \frac{m_0 - m_1}{m_0 - m_2} = \text{relative density of liquid,}$$

or $\frac{m_0 - m_1}{m_0 - m_2} \times \text{density of water} = \text{density of liquid.}$

Density of Copper Sulphate crystals

If a solid dissolves in water, such as a copper sulphate crystal for example, its density can be found by totally immersing it in a liquid in which it is insoluble. Copper sulphate can be weighed in paraffin oil, for example. Suppose the apparent weight is m_1, and the weight in air is m_0. Then

$$m_0 - m_1 = \text{upthrust in liquid} = V\rho,$$

where V is the volume of the solid and ρ is the density of the liquid.

$$\therefore V = \frac{m_0 - m_1}{\rho}.$$

$$\therefore \text{density of solid} = \frac{\text{mass}}{\text{volume}} = \frac{m_0}{V} = \frac{m_0}{m_0 - m_1} \cdot \rho.$$

The density, ρ, of the liquid can be found by means of a specific gravity bottle, for example. Thus knowing m_0 and m_1, the density of the solid can be calculated.

Relative Density of Cork

If a solid floats in water, cork for example, its density can be found by attaching a brass weight or 'sinker' to it so that both solids become totally immersed in water. The apparent weight (m_1) of the sinker and cork together is then obtained. Suppose m_2 is the weight of the sinker in air, m_3 is the weight of the sinker alone in water, and m_0 is the weight of the cork in air.

Then
$$m_2 - m_3 = \text{upthrust on sinker in water.}$$
$$\therefore m_0 + m_2 - m_1 - (m_2 - m_3) = \text{upthrust on cork in water}$$
$$= m_0 - m_1 + m_3$$
$$\therefore \text{relative density of cork} = \frac{m_0}{m_0 - m_1 + m_3}.$$

Flotation

When an object *floats* in a liquid, the upthrust on the object must be equal to its weight for equilibrium. Cork has a density of about 0·25 g cm^{-3}, so that 100 cm^3 of cork has a mass of 25 g. In water, then, cork sinks until the upthrust is 25 gf. Now from Archimedes' Principle, 25 gf is the weight of water displaced. Thus the cork sinks until 25 cm^3 of its 100 cm^3 volume is immersed. The fraction of the volume immersed is hence equal to the relative density.

Ice has a density of about 0·9 g cm^{-3}. A block of ice therefore floats in water with about $\frac{9}{10}$ths of it immersed.

Hydrometer

Hydrometers use the principle of flotation to measure density or relative density. Fig. 4.23. Since they have a constant weight, the upthrust when they float in a liquid is always the same. Thus in a liquid of density 1·0 g cm^{-3}, a hydrometer of 20 gf will sink until 20 cm^3 is immersed. In a liquid of density 2·0 g cm^{-3}, it will sink until only 10 cm^3 is immersed. The density or relative density readings hence increase in a *downward* direction, as shown in Fig. 4.23.

Practical hydrometers have a weighted end M for stability, a wide bulb to produce sufficient upthrust to counterbalance the weight, and a narrow stem BL for sensitivity. If V is the whole volume of the hydrometer in Fig. 4.22, a is the area of the stem and y is the length not immersed in a liquid of

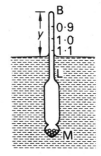

Fig. 4.23
Hydrometer

density ρ, then

$$\text{upthrust} = \text{wt. of liquid displaced} = (V-ay)\rho = w,$$

where w is the weight of the hydrometer.

EXAMPLE

An ice cube of mass 50 g floats on the surface of a strong brine solution of volume 200·0 cm³ inside a measuring cylinder. Calculate the level of the liquid in the measuring cylinder (i) before and (ii) after all the ice is melted. (iii) What happens to the level if the brine is replaced by 200·0 cm³ water and 50 g of ice is again added? (Assume density of ice, brine = 900, 1100 kg m⁻³ or 0·9, 1·1 g cm⁻³.)

(i) Floating ice displaces 50 g of brine since upthrust equals weight of ice.

$$\therefore \text{volume displaced} = \frac{\text{mass}}{\text{density}} = \frac{50}{1 \cdot 1} = 45 \cdot 5 \text{ cm}^3.$$

$\therefore$ level on measuring cylinder = 245·5 cm³.

(ii) 50 g of ice forms 50 g of water when all of it is melted.

$\therefore$ level on measuring cylinder *rises* to 250·0 cm³.

(iii) *Water*. Initially, volume of water displaced = 50 cm³, since upthrust = 50 g.

$\therefore$ level on cylinder = 250·0 cm³.

If 1 g of ice melts, volume displaced is 1 cm³ less. But volume of water formed is 1 cm³. Thus the net change in water level is zero. Hence the water level remains unchanged as the ice melts.

Accelerated liquid

Consider a solid S of mass m suspended from a string attached to a spring-balance and totally immersed in a liquid. Fig. 4.24 (i). The forces acting on the *solid* are its weight mg, the upthrust U of the liquid and the tension T of the spring. For equilibrium, $T = mg - U$. If we imagine S hollowed out and replaced

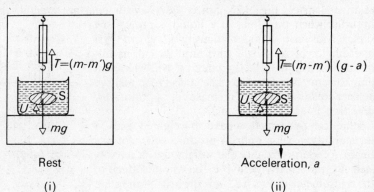

Fig. 4.24 Liquid in motion

by liquid, then for equilibrium of the *liquid*, $U = m'g$, where m' is the mass of liquid displaced by the solid. Observe that Archimedes' Principle applies to the liquid, which is at rest. Hence, from previous,

$$T = mg - U = mg - m'g = (m - m')g \qquad \text{(i)}$$

Now suppose the spring-balance and the liquid container are attached to the lift so that the solid S is totally immersed, as shown in Fig. 4.24 (ii). If the lift is given a *downward acceleration a*, then, for motion of the mass m of the solid,

$$mg - U_1 - T = ma \qquad \text{(ii)}$$

where U_1 is the new upthrust of the liquid. For motion of the *liquid* of mass m' which we imagine displacing S,

$$m'g - U_1 = m'a \qquad \text{(iii)}$$

Thus $U_1 = m'(g - a)$.

If Archimedes' Principle were true, the upthrust U, would be equal to $m'g$, as previously. Thus the Principle does *not* apply to accelerated liquids in which objects are immersed. In fact, from (iii), if there is free fall, so that $a = g$, the upthrust U_1, is *zero*. 'Weightlessness', previously discussed on p. 74, occurs here because the accelerated liquid effectively experiences no gravitational force when $a = g$.

From (ii) and (iii), it follows that

$$T = m(g - a) - U_1 = m(g - a) - m'(g - a) = (m - m')(g - a).$$

Comparing the tension with that in (i), it follows that the reading on the spring-balance is *reduced* for downward acceleration. In free fall, $a = g$ and $T = 0$.

With the lift undergoing an upward acceleration a, similar analysis shows that $T = (m - m')(g + a)$, so that the tension is now increased. This result is left to the reader to prove.

Similar considerations show that if a mercury barometer is in a lift moving down, the effective reduction of 'g' in $h\rho g$ will produce a rise in level to counterbalance atmospheric pressure. In free fall the mercury column will fill the barometer tube.

Fluids in Motion. Streamlines and velocity

A stream or river flows slowly when it runs through open country and faster through narrow openings or constrictions. As shown shortly, this is due to the fact that water is practically an incompressible fluid, that is, changes of pressure cause practically no change in fluid density at various parts.

Fig. 4.25 shows a tube of water flowing steadily between X and Y, where X has a bigger cross-sectional area A_1 than the part Y, of cross-sectional area A_2. The *streamlines* of the flow represent the directions of the velocities of the particles of the fluid and the flow is uniform or laminar (p. 204). Assuming the liquid is incompressible, then, if it moves from PQ to RS, the volume of liquid between P and R is equal to the volume between Q and S. Thus $A_1 l_1 = A_2 l_2$, where l_1 is PR and l_2 is QS,

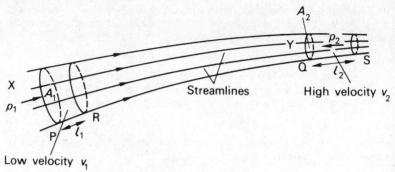

Fig. 4.25 Bernoulli's theorem

or $l_2/l_1 = A_1/A_2$. Hence l_2 is greater than l_1. Consequently the *velocity* of the liquid at the narrow part of the tube, where, it should be noted, the streamlines are closer together, is greater than at the wider part Y, where the streamlines are further apart. For the same reason, slow-running water from a tap can be made into a fast jet by placing a finger over the tap to narrow the exit.

Pressure and velocity. Bernoulli's Principle

About 1740, Bernoulli obtained a relation between the pressure and velocity at different parts of a moving incompressible fluid. If the viscosity is negligibly small, there are no frictional forces to overcome (p. 204). In this case the work done by the pressure difference per unit volume of a fluid flowing along a pipe steadily is equal to the gain of kinetic energy per unit volume plus the gain in potential energy per unit volume.

Now the work done by a pressure in moving a fluid through a distance = force × distance moved = (pressure × area) × distance moved = pressure × volume moved, assuming the area is constant at a particular place for a short time of flow. At the beginning of the pipe where the pressure is p_1, the work done per unit volume on the fluid is thus p_1; at the other end, the work done per unit volume by the fluid is likewise p_2. Hence the net work done *on* the fluid per unit volume $= p_1 - p_2$. The kinetic energy per unit volume $= \frac{1}{2}$ mass per unit volume × velocity². $= \frac{1}{2}\rho \times$ velocity², where ρ is the density of the fluid. Thus if v_2 and v_1 are the final and initial velocities respectively at the end and the beginning of the pipe, the kinetic energy gained per unit volume $= \frac{1}{2}\rho(v_2 - v_1)^2$. Further, if h_2 and h_1 are the respective heights measured from a fixed level at the end and beginning of the pipe, the potential energy gained per unit volume = mass per unit volume × g × $(h_2 - h_1)$ $= \rho g(h_2 - h_1)$.

STATIC BODIES, FLUIDS

Thus, from the conservation of energy,

$$p_1 - p_2 = \tfrac{1}{2}\rho(v_2^2 - v_1^2) + \rho g(h_2 - h_1)$$
$$\therefore p_1 + \tfrac{1}{2}\rho v_1^2 + \rho g h_1 = p_2 + \tfrac{1}{2}\rho v_2^2 + \rho g h_2$$
$$\therefore p + \tfrac{1}{2}\rho v^2 + \rho g h = \text{constant},$$

where p is the pressure at any part and v is the velocity there. Hence it can be said that, for streamline motion of an incompressible non-viscous fluid,

the sum of the pressure at any part plus the kinetic energy per unit volume plus the potential energy per unit volume there is always constant.

This is known as *Bernoulli's principle*.

Bernoulli's principle shows that at points in a moving fluid where the potential energy change $\rho g h$ is very small, or zero as in flow through a horizontal pipe, the pressure is low where the velocity is high; conversely, the pressure is high where the velocity is low. The principle has wide applications.

EXAMPLE

As a numerical illustration of the previous analysis, suppose the area of cross-section A_1 of X in Fig. 4.25 is 4 cm^2, the area A_2 of Y is 1 cm^2, and water flows past each section in laminar flow at the rate of 400 cm^3 s^{-1}. Then

at X, speed v_1 of water $= \dfrac{\text{vol. per second}}{\text{area}} = 100$ cm s^{-1} = 1 m s^{-1};

at Y, speed v_2 of water $= 400$ cm s^{-1} = 4 m s^{-1}.

The density of water, $\rho = 1000$ kg m^{-3}.

$$\therefore p = \tfrac{1}{2}\rho(v_2^2 - v_1^2) = \tfrac{1}{2} \times 1000 \times (4^2 - 1^2) = 7\cdot5 \times 10^3 \text{ N m}^{-2}.$$

If h is in metres, $\rho = 1000$ kg m^{-3} for water, $g = 9\cdot8$ m s^{-2}, then, from $h\rho g$,

$$h = \frac{7\cdot5 \times 10^3}{1000 \times 9\cdot8} = 0\cdot77 \text{ m (approx.)}.$$

The pressure head h is thus equivalent to 0·77 m of water.

Applications of Bernoulli's Principle

1. A suction effect is experienced by a person standing close to the platform at a station when a fast train passes. The fast-moving air between the person and train produces a decrease in pressure and the excess air pressure on the other side pushes the person towards the train.

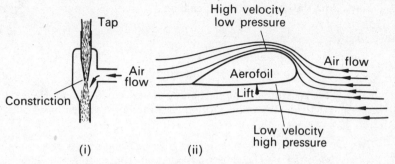

Fig. 4.26   Fluid velocity and pressure

2. *Filter pump.* A filter pump has a narrow section in the middle, so that a jet of water from the tap flows faster here. Fig. 4.26 (i). This causes a drop in pressure near it and air therefore flows in from the side tube to which a vessel is connected. The air and water together are expelled through the bottom of the filter pump.

3. *Aerofoil lift.* The curved shape of an aerofoil creates a faster flow of air over its top surface than the lower one. Fig. 4.26 (ii). This is shown by the closeness of the streamlines above the aerofoil compared with those below. From Bernoulli's principle, the pressure of the air below is greater than that above, and this produces the lift on the aerofoil.

4. *Flow of liquid from wide tank.* Suppose a liquid flows through a hole H at the bottom of a wide tank, as shown in Fig. 4.27. Assuming negligible viscosity and streamline flow at a small distance from the hole, which is an approximation, Bernoulli's theorem can be applied. At the top X of the liquid in the tank, the pressure is atmospheric, say B, the height measured from a fixed level such as the hole H is h, and the kinetic energy is negligible if the tank is wide so that the level falls very slowly. At the bottom, Y, near H, the pressure is again B, the height above H is now zero, and the kinetic energy is $\frac{1}{2}\rho v^2$, where ρ is the density and v is the velocity of emergence of the liquid. Thus, from Bernoulli's Principle,

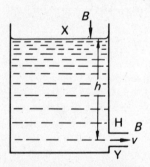

Fig. 4.27   Torricelli's theorem

$$B + \rho hg = B + \tfrac{1}{2}\rho v^2$$
$$\therefore v^2 = 2gh$$

Thus the velocity of the emerging liquid is the same as that which would be obtained if it fell freely through a height h; and this is known as *Torricelli's theorem*. In practice the velocity is less than that given by $\sqrt{2gh}$ owing to viscous forces, and the lack of streamline flow must also be taken into account.

EXERCISES 4

What are the missing words in the statements 1–6?

1. In SI units, the moment or torque of a couple is measured in . . .

2. In stable equilibrium, when an object is slightly displaced its centre of gravity . . .

3. When an object is in equilibrium under the action of three non-parallel forces, the three forces must . . . one point.

4. The component of a force F in a direction inclined to it at an angle θ is . . .

5. The sensitivity of a beam balance depends on the depth of the . . . below the fulcrum.

6. When an object floats, the weight of fluid displaced is equal to the . . .

7. In laminar flow of non-viscous fluid along a pipe, at regions of high pressure the . . . is low.

Which of the following answers, A, B, C, D or E, do you consider is the correct one in the statements 8–10?

8. If a cone is balanced on its apex on a horizontal table and then slightly displaced, the potential energy of the cone is then *A* increased, *B* decreased, *C* constant, *D* a minimum, *E* a maximum.

9. If a hydrometer of mass 20 g and volume 30 cm^3 has a graduated stem of 1 cm^2, and floats in water, the exposed length of stem is *A* 30 cm, *B* 25 cm, *C* 20 cm, *D* 10 cm, *E* 1 cm.

10. In laminar flow of a non-viscous fluid along a horizontal pipe, the work per second done by the pressure at any section is equal to *A* the pressure, *B* the volume per second there, *C* pressure × volume per second there, *D* pressure × volume, *E* pressure × area of cross-section.

11. A flat plate is cut in the shape of a square of side 20·0 cm, with an equilateral triangle of side 20·0 cm adjacent to the square. Calculate the distance of the centre of mass from the apex of the triangle.

12. The foot of a uniform ladder is on a rough horizontal ground, and the top rests against a smooth vertical wall. The weight of the ladder is 400 N, and a man weighing 800 N stands on the ladder one-quarter of its length from the bottom. If the inclination of the ladder to the horizontal is 30°, find the reaction at the wall and the total force at the ground.

13. A rectangular plate ABCD has two forces of 100 N acting along AB and DC in opposite directions. If AB = 3 m, BC = 5 m, what is the moment of the couple acting on the plate? What forces acting along BC and AD respectively are required to keep the plate in equilibrium?

14. A hollow metal cylinder 2 m tall has a base of diameter 35 cm and is filled with water to a height of (i) 1 m, (ii) 50 cm. Calculate the distance of the centre of gravity in metre from the base in each case if the cylinder has no top. (Metal weighs 20 kg m^{-2} of surface. Assume $\pi = 22/7$.)

15. A trap-door 120 cm by 120 cm is kept horizontal by a string attached to the mid-point of the side opposite to that containing the hinge. The other end of the string is tied to a point 90 cm vertically above the hinge. If the trap-door weight is 50 N, calculate the tension in the string and the reaction at the hinge.

16. Two smooth inclined planes are arranged with their lower edges in contact; the angles of inclination of the plane to the horizontal are 30°, 60° respectively, and the surfaces of the planes are perpendicular to each other. If a uniform rod rests in the principal section of the planes with one end on each plane, find the angle of inclination of the rod to the horizontal.

17. Describe and give the theory of an accurate beam balance. Point out the factors which influence the sensitivity of the balance. Why is it necessary, in very accurate weighing, to take into account the pressure, temperature, and humidity of the atmosphere? (*O. & C.*)

18. Summarise the various conditions which are being satisfied when a body remains in equilibrium under the action of three non-parallel forces.

A wireless aerial attached to the top of a mast 20 m high exerts a horizontal force upon it of 600 N. The mast is supported by a stay-wire running to the ground from a point 6 m below the top of the mast, and inclined at 60° to the horizontal. Assuming that the action of the ground on the mast can be regarded as a single force, draw a diagram of the forces acting on the mast, and determine by measurement or by calculation the force in the stay-wire. (*C.*)

19. The beam of a balance weighs 150 g and its moment of inertia is 5×10^{-4} kg m^2. Each arm of the balance is 10 cm long. When set swinging the beam makes one complete oscillation in 6 seconds. How far is the centre of gravity of the beam below its point of support, and through what angle would the beam be deflected by a weight of 1 milligram placed in one of the scale pans? (*C.*)

20. Under what conditions is a body said to be in equilibrium? What is meant by (*a*) *stable equilibrium* and (*b*) *unstable equilibrium*? Give one example of each.

A pair of railway carriage wheels, each of radius r, are joined by a thin axle; the mass of the whole is m. A light arm of length $l (< r)$ is attached perpendicularly to the axle and the free end of the arm carries a point mass M. The wheels rest, with the axle horizontal, on rails which are laid down a slope inclined at an angle ϕ to the horizontal. Show that, provided that ϕ is not too large and that the wheels do not slip on the rails, there are two values of the angle θ that the arm makes with the horizontal when the system is in equilibrium, and find these values of θ. Discuss whether, in each case, the equilibrium is stable or unstable. (*O. & C.*)

21. Give a labelled diagram to show the structure of a beam balance. Show if the knife-edges are collinear the sensitivity is independent of the load. Discuss other factors which then determine the sensitivity.

A body is weighed at a place on the equator, both with a beam balance and a very sensitive spring balance, with identical results. If the observations are repeated at a place near one of the poles, using the same two instruments, discuss whether identical results will again be obtained. (*L*.)

22. Three forces in one plane act on a rigid body. What are the conditions for equilibrium?

The plane of a kite of mass 6 kg is inclined to the horizon at 60°. The resultant thrust of the air on the kite acts at a point 25 cm above its centre of gravity, and the string is attached at a point 30 cm above the centre of gravity. Find the thrust of the air on the kite, and the tension in the string. (*C*.)

23. In what circumstances is a physical system in equilibrium? Distinguish between stable, unstable and neutral equilibria.

Discuss the stability of the equilibrium of a uniform rough plank of thickness *t*, balanced horizontally on a rough cylindrical-fixed log of radius *r*, it being assumed that the axes of plank and log lie in perpendicular directions. (*N*.)

24. State the conditions of equilibrium for a body subjected to a system of coplanar parallel forces and briefly describe an experiment which you could carry out to verify these conditions.

Show how the equilibrium of a beam balance is achieved and discuss the factors which determine its sensitivity. Explain how the sensitivity of a given balance may be altered and why, for a particular adjustment, the sensitivity may be practically independent of the mass in the balance pans. Why is it inconvenient in practice to attempt to increase the sensitivity of a given balance beyond a certain limit? (*O. & C*.)

Fluids

25. An alloy of mass 588 g and volume 100 cm^3 is made of iron of relative density 8·0 and aluminium of relative density 2·7. Calculate the proportion (i) by volume, (ii) by weight of the constituents of the alloy.

26. A string supports a solid iron object of mass 180 g totally immersed in a liquid of density 800 kg m^{-3}. Calculate the tension in the string if the density of iron is 8000 kg m^{-3}.

27. A hydrometer floats in water with 6·0 cm of its graduated stem unimmersed, and in oil of relative density 0·8 with 4·0 cm of the stem unimmersed. What is the length of stem unimmersed when the hydrometer is placed in a liquid of relative density 0·9?

28. An alloy of mass 170 g has an apparent weight of 95 g in a liquid of density 1·5 g cm^{-3}. If the two constituents of the alloy have relative densities of 4·0 and 3·0 respectively, calculate the proportion by volume of the constituents in the alloy.

29. State the principle of Archimedes and use it to derive an expression for the resultant force experienced by a body of weight *W* and density σ when it is totally immersed in a fluid of density ρ.

A solid weighs 237·5 g in air and 12·5 g when totally immersed in a liquid of specific gravity 0·9. Calculate (*a*) the relative density of the solid, (*b*) the relative density of a liquid in which the solid would float with one-fifth of its volume exposed above the liquid surface. (*L*.)

30. Distinguish between *mass* and *weight*. Define *density*.

Describe and explain how you would proceed to find an accurate value for the density of gold, the specimen available being a wedding ring of pure gold.

What will be the reading of (*a*) a mercury barometer, (*b*) a water barometer, when the atmospheric pressure is 10^5 N m^{-2}? The density of mercury may be taken as 13600 kg m^{-3} and the pressure of saturated water vapour at room temperature as 13 mm of mercury. (*L.*)

31. Describe an experiment which demonstrates the difference between laminar and turbulent flow in a fluid.

A straight pipe of uniform radius R is joined, in the same straight line, to a narrower pipe of uniform radius r. Water (which may be assumed to be incompressible) flows from the wider into the narrower pipe. The velocity of flow in the wider pipe is V and in the narrower pipe is v. By equating work done against fluid pressures with change of kinetic energy of the water, show that the hydrostatic pressure is lower where the velocity of flow is higher.

Describe and explain **one** practical consequence or application of this difference in pressures. (*O. & C.*)

32. Describe some form of barometer used for the accurate measurement of atmospheric pressure, and point out the corrections to be applied to the observation.

Obtain an expression for the correction to be applied to the reading of a mercurial barometer when the reading is made at a temperature other than 0°C. (*L.*)

33. State the principle of Archimedes, and discuss its application to the determination of density by means of a common hydrometer. Why is this method essentially less accurate than the density bottle?

A common hydrometer is graduated to read relative densities from 0·8 to 1·0. In order to extend its range a small weight is attached to the stem, above the liquid, so that the instrument reads 0·8 when floating in water. What will be the relative density of the liquid corresponding to the graduation 1·0? (*O. & C.*)

34. A hydrometer consists of a bulb of volume V and a uniform stem of volume v per cm of its length. It floats upright in water so that the bulb is just completely immersed. Explain for what density range this hydrometer may be used and how you would determine the density of such liquids. Describe the graph which would be obtained by plotting the reciprocal of the density against the length of the stem immersed.

A hydrometer such as that described sinks to the mark 3 on the stem, which is graduated in cm, when it is placed in a liquid of density 0·95 g cm^{-3}. If the volume per cm of the stem is 0·1 cm^3, find the volume of the bulb. (*L.*)

35. A straight rod of length l, small cross sectional area a and of material density ρ is supported by a thread attached to its upper end. Initially the rod hangs in a vertical position over a liquid of density σ and then is lowered until it is partially submerged. Derive and discuss the equilibrium conditions of the rod neglecting surface tension. (*N.*)

Chapter 5

SURFACE TENSION

Intermolecular Forces

THE forces which exist between molecules can explain many of the bulk properties of solids, liquids and gases. These intermolecular forces arise from two main causes:

(1) The *potential energy* of the molecules, which is due to interactions with surrounding molecules (this is principally electrical, not gravitational, in origin).

(2) The *thermal energy* of the molecules—this is the kinetic energy of the molecules and depends on the temperature of the substance concerned.

We shall see later that the particular state or phase in which matter appears—that is, solid, liquid or gas—and the properties it then has, are determined by the relative magnitudes of these two energies.

Potential energy and Force

In bulk, matter consists of numerous molecules. To simplify the situation, Fig. 5.1 shows the variation of the potential energy V between two molecules at a distance r apart.

Along the part BCD of the curve, the potential energy V is negative. Along the part AB, the potential energy V is positive. The force between the molecules is always given by $F = -dV/dr = -$ potential gradient. Along CD the force is *attractive* and it decreases with distance r according to an inverse-square law. Along ABC, the force is *repulsive*. Fig. 5.1 shows the variation of F with r.

At C, the minimum potential energy point of the curve, the molecules would be at their normal distance apart in the absence of thermal energy. The equilibrium distance OM, r_0, is of the order 2 or 3×10^{-10} m (2 or 3 Å) for a solid. At this distance apart, the attractive and repulsive forces balance each other. If the molecules are closer, $(r < r_0)$, they would repel each other. If they are further apart, $(r > r_0)$, they attract each other.

Phases or States of Matter

The molecules in a *solid* are said to be in a 'condensed' phase or state. Their thermal energy is then relatively low compared with their

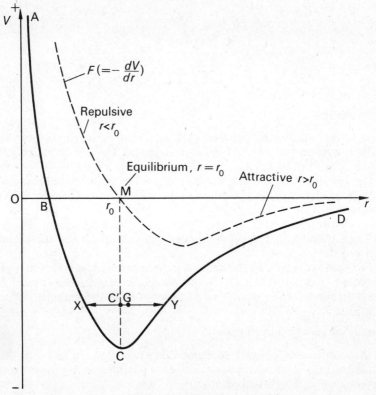

Fig. 5.1 Molecular potential energy and force

potential energy V and the molecules are 'bound' to each other. They may now vibrate about C, the minimum of the curve in Fig. 5.1.

When the thermal energy increases by an amount corresponding to CC' in Fig. 5.1, the molecule can then oscillate between the limits corresponding to X and Y. From the graph of force, F, it can be seen that when the molecule is on the left of C' it experiences a greater force towards it than when on the right. Consequently the molecule returns quicker to C'. Thus the *mean position* G is on the right of C'. This corresponds to a mean separation of molecules which is *greater* than r_0. Thus the solid expands when its thermal energy is increased.

As the thermal energy increases further, at some particular temperature the molecules are able to move comparatively freely relative to neighbouring molecules. The solid then loses its rigid form and becomes a *liquid*. The molecules in the liquid constantly exchange places with other molecules, whereas in a solid the neighbours of a particular molecule remain unchanged. Further, the molecules of a liquid have translational as well as vibrational energy, that is, they move about constantly through the liquid, whereas molecules of a solid have vibrational energy only.

As the temperature of the liquid rises, the thermal energy of the molecules further increases. The average distance between the molecules then also increases and so their mean potential energy approaches zero, as can be seen from Fig. 5.1. At some stage the increased thermal energy enables the molecules to completely break the bonds of attraction which keep them in a liquid state. The molecules then have little or no interaction and now form a *gas*. At normal pressures the forces of attraction between the gas molecules are comparatively very small and the molecules move about freely inside the volume they occupy. Gas molecules which are monatomic such as helium have translational energy only. Gas molecules such as oxygen or carbon dioxide, with two or more atoms, have rotational and vibrational energies in addition to translational energy.

Gases

At normal pressure, permanent gases such as air or oxygen obey Boyle's law, pV = constant, to a very good approximation. Now in the absence of attractive forces between the molecules, and assuming their actual volume is negligibly small, the kinetic theory of gases shows that Boyle's law is obeyed by this ideal gas. Consequently, the attractive forces between the gas molecules at normal pressure are unimportant. They increase appreciably when the gas is at high pressure as the molecules are then on the average very much closer.

In the bulk of the gas, the resultant force of attraction between a particular molecule and those all round it is zero when averaged over a period. Molecules which strike the wall of the containing vessel, however, are retarded by an unbalanced force due to molecules behind them. The observed pressure p of a gas is thus *less* than the pressure in the ideal case, when the attractive forces due to molecules is zero.

Van der Waals derived an expression for this pressure 'defect'. He considered that it was proportional to the product of the number of molecules per second striking unit area of the wall and the number per unit volume behind them, since this is a measure of the force of attraction. For a given volume of gas, both these numbers are proportional to the *density* of the gas. Consequently the pressure defect, p_1 say, is proportional to $\rho \times \rho$ or ρ^2. For a fixed mass of gas, $\rho \propto 1/V$, where V is the volume. Thus $p_1 = a/V^2$, where a is a constant for the particular gas. Taking into account the attractive forces between the molecules, it follows that, if p is the observed pressure, the gas pressure in the bulk of the gas $= p + a/V^2$.

The attraction of the walls on the molecules arriving there is to increase their velocity from v say to $v + \Delta v$. Immediately after rebounding from the walls, however, the force of attraction decreases the velocity to v again. Thus the attraction of the walls has no net effect on the momentum change due to collision. Likewise, the increase in momentum of the walls due to their attraction by the molecules arriving is lost after the molecules rebound.

The effect of the volume actually occupied by all the molecules is represented by a constant b, so that the volume of the space in which they move is not V but $(V-b)$. The magnitude of b is not the actual volume of the molecules, as if they were swept into one corner of the space, since they are in constant motion. b has been estimated to be about four times the actual volume.

Surface Tension.

We now consider in detail a phenomenon of a liquid surface called *surface tension*. As we shall soon show, surface tension is due to intermolecular attraction.

It is a well-known fact that some insects, for example a water-carrier, are able to walk across a water surface; that a drop of water may remain suspended for some time from a tap before falling, as if the water particles were held together in a bag; that mercury gathers into small droplets when spilt; and that a dry steel needle may be made, with

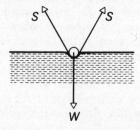

FIG. 5.2 Needle floating on water

care, to float on water, Fig. 5.2. These observations suggest that *the surface of a liquid acts like an elastic skin covering the liquid or is in a state of tension*. Thus forces S in the liquid support the weight W of the needle, as shown in Fig. 5.2.

Energy of Liquid Surface. Molecular theory

The fact that a liquid surface is in a state of tension can be explained by the intermolecular forces discussed on p. 146. In the bulk of the liquid, which begins only a few molecular diameters downwards from the surface, a particular molecule such as A is surrounded by an equal number of molecules on all sides. This can be seen by drawing a sphere round A. Fig. 5.3. The average distance apart of the molecules is such that the attractive forces balance the repulsive forces (p. 145). Thus the average intermolecular force between A and the surrounding molecules is zero. Fig. 5.3.

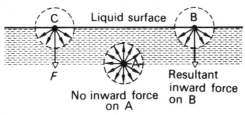

FIG. 5.3 Molecular forces in liquid

Consider now a molecule such as C or B in the surface of the liquid. There are very few molecules on the vapour side above C or B compared with the liquid below, as shown by drawing a sphere round C or B. Thus if C is displaced very slightly upward, a resultant attractive force F on C, due to the large number of molecules below C, now has to be overcome. It follows that if all the molecules in the surface were removed to infinity, a definite amount of work is needed. *Consequently molecules in the surface have potential energy.* A molecule in the bulk of the liquid forms bonds with more neighbours than one in the surface. Thus bonds must be broken, i.e. work must be done, to bring a molecule into the surface. Molecules in the surface of the liquid hence have more potential energy than those in the bulk.

Surface area. Shape of drop

The potential energy of any system in stable equilibrium is a minimum. Thus under surface tension forces, the area of a liquid surface will have the least number of molecules in it, that is, the surface area of a given volume of liquid is a minimum. Mathematically, it can be shown that the shape of a given volume of liquid with a minimum surface area is a *sphere*.

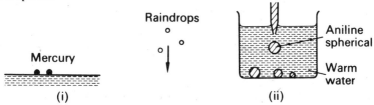

FIG. 5.4 Liquid drops

This is why raindrops, and small droplets of mercury, are approximately spherical in shape. Fig. 5.4 (i). To eliminate completely the effect of gravitational forces, Plateau placed a drop of oil in a mixture of alcohol and water of the same density. In this case the weight of the drop is counterbalanced by the upthrust of the surrounding liquid. He then observed that the drop was a perfect sphere. Plateau's 'spherule' experiment can be carried out by warming water in a beaker and then carefully introducing aniline with the aid of a pipette. Fig. 5.4 (ii). At room temperature the density of aniline is slightly greater than water. At a higher temperature the densities of the two liquids are roughly the same and the aniline is then seen to form *spheres*, which rise and fall in the liquid.

A soap bubble is spherical because its weight is extremely small and the liquid shape is then mainly due to surface tension forces. Although the density of mercury is high, small drops of mercury are spherical. The ratio of surface area ($4\pi r^2$) to weight (or volume, $4\pi r^3/3$) of a sphere is proportional to the ratio r^2/r^3, or to $1/r$. Thus the smaller the radius, the greater is the influence of surface tension forces compared to the weight. Large mercury drops, however, are flattened on top. This time the effect of gravity is relatively greater. The shape of the drop conforms to the principle that the sum of the gravitational potential energy and the surface energy must be a minimum, and so the centre of gravity moves down as much as possible.

Lead shot is manufactured by spraying lead from the top of a tall tower. As they fall, the small drops form spheres under the action of surface tension forces.

Surface tension definition. Units, dimensions

Since the surface of a liquid acts like an elastic skin, the surface is in a state of tension. A blown-up football bladder has a surface in a state of tension. This is a very rough analogy because the surface tension of a bladder increases as the surface area increases, whereas the surface tension of a liquid is independent of surface area. Any line in the bladder surface is then acted on by two equal and opposite forces, and if the bladder is cut with a knife the rubber is drawn away from the incision by the two forces present.

R. C. Brown and others have pointed out that molecules in the surface of a liquid have probably a less dense packing than those in the bulk of the liquid, as there are fewer molecules in the surface when its area is a minimum. The average separation between molecules in the surface are then slightly greater than those inside. On average, then, the force between neighbouring molecules in the surface are attractive (see p. 145). This would explain the existence of surface tension.

The *surface tension*, γ, of a liquid, sometimes called the *coefficient* of *surface tension*, is defined as *the force per unit length acting in the surface at right angles to one side of a line drawn in the surface*. In Fig. 5.5 AB represents a line 1 m long. The unit of γ is *newton metre^{-1}* (N m^{-1}).

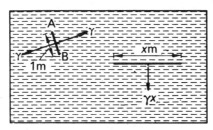

FIG. 5.5 Surface tension

The 'magnitude' of γ depends on the temperature of the liquid and on the medium on the other side of the surface. For water at 20°C in contact with air, $\gamma = 7\cdot26 \times 10^{-2}$ newton metre^{-1}. For mercury at 20°C in contact with air, $\gamma = 46\cdot5 \times 10^{-2}$ N m^{-1}. The surface tension of a water-oil (olive-oil) boundary is $2\cdot06 \times 10^{-2}$ N m^{-1}, and for a mercury-water boundary it is $42\cdot7 \times 10^{-2}$ N m^{-1}.

Since surface tension γ is a 'force per unit length', the dimensions of

$$\text{surface tension} = \frac{\text{dimensions of force}}{\text{dimensions of length}} = \frac{MLT^{-2}}{L}$$

$$= MT^{-2}.$$

We shall see later that surface tension can be defined also in terms of surface energy (p. 170).

Some surface tension phenomena

The effect of surface tension forces in a soap film can be demonstrated by placing a thread B carefully on a soap film formed in a metal ring A, Fig. 5.6 (i). The surface tension forces on both sides of the thread counterbalance, as shown in Fig. 5.6 (i). If the film enclosed by the thread is pierced, however, the thread is pulled out into a circle by the surface tension forces F at the junction of the air and soap-film, Fig. 5.6 (ii). Observe that the film has contracted to a minimum area.

Another demonstration of surface tension forces can be made by sprinkling light dust or lycopodium powder over the surface of water contained in a dish. If the middle of the water is touched with the end of a glass rod which had previously been dipped into soap solution, the powder is carried away to the sides by the water. The explanation lies in the fact that the surface tension of water is greater than that of a

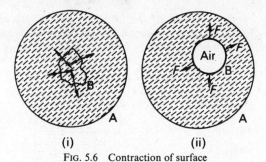

Fig. 5.6 Contraction of surface

soap-film (p. 157). The resultant force at the place where the rod touched the water is hence *away from* the rod, and thus the powder moves away from the centre towards the sides of the vessel.

A toy duck moves by itself across the surface of water when it has a small bag of camphor attached to its base. The camphor lowers the surface tension of the water in contact with it, and the duck is urged across the water by the resultant force on it.

Capillarity

When a capillary tube is immersed in water, and then placed vertically with one end in the liquid, observation shows that the water rises in the tube to a height above the surface. The narrower the tube, the greater is the height to which the water rises, Fig. 5.7 (i). See also p. 162). This phenomenon is known as *capillarity*, and it occurs when blotting-paper is used to dry ink. The liquid rises up the pores of the paper when it is pressed on the ink.

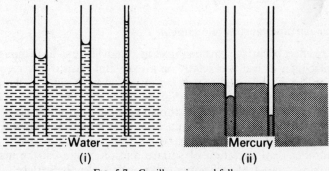

Fig. 5.7 Capillary rise and fall

When a capillary tube is placed inside mercury, however, the liquid is depressed *below* the outside level, Fig. 5.7 (ii). The depression increases as the diameter of the capillary tube decreases. See also p. 163.

Angle of Contact

In the case of water in a glass capillary tube, observation of the meniscus shows that it is hemispherical if the glass is clean, that is, the glass surface is tangential to the meniscus where the water touches it. In other cases where liquids rise in a capillary tube, the tangent BN to the liquid surface where it touches the glass may make an acute angle θ with the glass, Fig. 5.8 (i). The angle θ is known as the *angle of contact* between the liquid and the glass, and is always measured *through the liquid*. The angle of contact between two given surfaces varies largely with their freshness and cleanliness. The angle of contact between water and very clean glass is zero, but when the glass is not clean the angle of contact may be about 8° for example. The angle of contact between alcohol and very clean glass is zero.

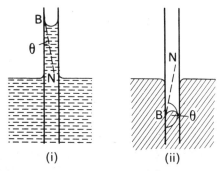

FIG. 5.8   Angle of contact

When a capillary tube is placed inside mercury, observation shows that the surface of the liquid is depressed in the tube and is convex upwards. Fig. 5.8 (ii). The tangent BN to the mercury at the point B where the liquid touches the glass thus makes an obtuse angle, θ, with the glass when measured through the liquid. We shall see later (p. 160) that a liquid will rise in a capillary tube if the angle of contact is acute, and that a liquid will be depressed in the tube if the angle of contact is obtuse. For the same reason, clean water spreads over, or 'wets', a clean glass surface when spilt on it, Fig. 5.9 (i); the angle of contact is zero. On the other hand, mercury gathers itself into small pools or globules when spilt on glass, and does not 'wet' glass, Fig. 5.9 (ii). The angle of contact is obtuse. See also p. 173.

FIG. 5.9   Water and mercury on glass

The difference in behaviour of water and mercury on clean glass can be explained in terms of the attraction between the molecules of these substances. It appears that the force of *cohesion* between two molecules of water is less than the force of *adhesion* between a molecule of water and a molecule of glass; and thus water spreads over glass. On the other hand, the force of cohesion between two molecules of mercury is greater than the force of adhesion between a molecule of mercury and a molecule of glass; and thus mercury gathers in pools when spilt on glass.

Angle of Contact measurement

The angle of contact can be found by means of the method outlined in Fig. 5.10 (i), (ii).

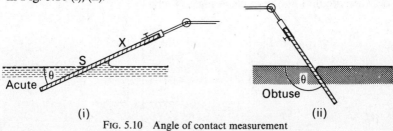

FIG. 5.10 Angle of contact measurement

A plate X of the solid is placed at varying angles to liquid until the surface S appears to be *plane* at X. The angle θ made with the liquid surface is then the angle of contact. For an obtuse angle of contact, a similar method can be adopted. In the case of mercury and glass, a thin plane mirror enables the liquid surface to be seen by reflection. For a freshly-formed mercury drop in contact with a clean glass plate, the angle of contact is 137°.

Measurement of Surface Tension by Capillary Tube Method

Theory. Suppose γ is the magnitude of the surface tension of a liquid such as water, which rises up a clean glass capillary tube and has an angle of contact zero. Fig. 5.11 shows a section of the meniscus M at B, which is a hemisphere. Since the glass AB is a tangent to the liquid, the surface tension forces, which act along the boundary of the liquid with the air, act vertically downwards on the glass. By the law of action and reaction, the glass exerts an equal force in an upward direction on the liquid. Now surface tension, γ, is the force per unit length acting in the surface of the liquid, and the length of liquid in contact with the glass is $2\pi r$, where r is the radius of the capillary tube.

$$\therefore 2\pi r \times \gamma = \text{upward force on liquid} \quad . \quad . \quad (1)$$

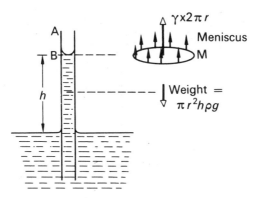

FIG. 5.11 Rise in capillary tube—theory

If γ is in newton metre^{-1} and r is in metres, then the upward force is in *newtons*.

This force supports the weight of a column of height h above the outside level of liquid. The volume of the liquid $= \pi r^2 h$, and thus the mass, m, of the liquid column $=$ volume $\times$ density $= \pi r^2 h \rho$, where ρ is the density. The *weight* of the liquid $= mg = \pi r^2 h \rho g$.

If ρ is in kg m^{-3}, r and h in metres, and $g = 9.8$ m s^{-2}, then $\pi r^2 h \rho g$ is in *newtons*.

From (1), it now follows that

$$\therefore 2\pi r \gamma = \pi r^2 h \rho g$$

$$\therefore \gamma = \frac{rh\rho g}{2} \quad . \quad . \quad . \quad . \quad (2)$$

If $r = 0.2$ mm $= 0.2 \times 10^{-3}$ m, $h = 6.6$ cm for water $= 6.6 \times 10^{-2}$ m, and $\rho = 1000$ kg m^{-3}, then

$$\gamma = \frac{0.2 \times 10^{-3} \times 6.6 \times 10^{-2} \times 1000 \times 9.8}{2} = 6.5 \times 10^{-2} \text{ N m}^{-1}$$

In deriving this formula for γ it should be noted that we have (i) assumed the glass to be a tangent to the liquid surface meeting it, (ii) neglected the weight of the small amount of liquid above the bottom of the meniscus at B, Fig. 5.11.

Experiment. In the experiment, the capillary tube C is supported in a beaker Y, and a pin P, bent at right angles at two places, is attached to C by a rubber band, Fig. 5.12. P is adjusted until its point just touches the horizontal level of the liquid in the beaker. A travelling microscope is now focussed on to the meniscus M in C, and then it is focussed on to the point of P, the beaker being removed for this observation. In this way the height h of M above the level in the beaker is determined. The

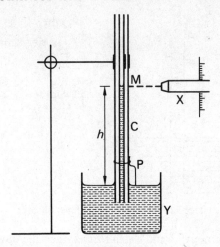

Fig. 5.12 Surface tension by capillary rise

radius of the capillary at M can be found by cutting the tube at this place and measuring the diameter by the travelling microscope; or by measuring the length, l, and mass, m, of a mercury thread drawn into the tube, and calculating the radius, r, from the relation $r = \sqrt{m/\pi l \rho}$, where ρ is the density of mercury. The surface tension γ is then calculated from the formula $\gamma = rh\rho g/2$. Its magnitude for water at 15°C is 7.33×10^{-2} N m^{-1}.

Measurement of Surface Tension by Microscope Slide

Besides the capillary tube method, the surface tension of water can be measured by weighing a microscope slide in air, and then lowering it until it just meets the surface of water, Fig. 5.13. The surface tension force acts vertically downward round the boundary of the slide, and pulls the slide down. If a and b are the length and thickness of the slide,

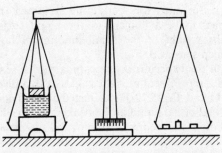

Fig. 5.13 Surface tension by microscope slide

then, since γ is the force per unit length in the liquid surface and $(2a+2b)$ is the length of the boundary of the slide, the downward force $= \gamma(2a+2b)$. If the mass required to counterbalance the force is m, then

$$\gamma(2a+2b) = mg,$$

$$\therefore \gamma = \frac{mg}{2a+2b}.$$

If $m = 0{\cdot}88$ g $= 0{\cdot}88 \times 10^{-3}$ kg, $a = 6{\cdot}0$ cm, $b = 0{\cdot}2$ cm, then:

$$\gamma = \frac{0{\cdot}88 \times 10^{-3} \text{ (kg)} \times 9{\cdot}8 \text{ (m s}^{-2})}{2 \times (6+0{\cdot}2) \times 10^{-2} \text{ (m)}} = 7{\cdot}0 \times 10^{-2} \text{ N m}^{-1}.$$

Surface Tension of a Soap Solution

The surface tension of a soap solution can be found by a similar method. A soap-film is formed in a three-sided metal frame ABCD, and the apparent weight is found, Fig. 5.14. When the film is broken by piercing it, the decrease in the apparent weight, mg, is equal to the

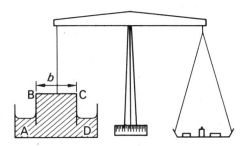

FIG. 5.14 Surface tension of soap film

surface tension force acting downwards when the film existed. This is equal to $2\gamma b$, where $b = $ BC, since the film has *two* sides.

$$\therefore 2\gamma b = mg,$$

$$\therefore \gamma = \frac{mg}{2b}.$$

It will be noted that the surface tension forces on the sides AB, CD of the frame act horizontally, and their resultant is zero.

A soap film can be supported in a vertical rectangular frame but a film of water can not. This is due to the fact that the soap drains downward in a vertical film, so that the top of the film has a lower concentration of soap than the bottom. The surface tension at the top is thus *greater* than at the bottom (soap diminishes the surface tension of pure water). The upward pull on the film by the top bar is hence greater than the downward pull on the film by the lower bar. The net upward pull supports the weight of the film. In the case of pure water, however, the surface tension would be the same at the top and bottom, and hence there is no net force in this case to support a water film in a rectangular frame.

Pressure Difference in a Bubble or Curved Liquid Surface

As we shall see presently, the magnitude of the curvature of a liquid, or of a bubble formed in a liquid, is related to the surface tension of the liquid.

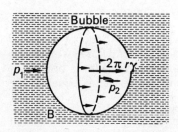

FIG. 5.15 Excess pressure in bubble

Consider a bubble formed inside a liquid, Fig. 5.15. If we consider the equilibrium of *one half*, B, of the bubble, we can see that the surface tension force on B plus the force on B due to the external pressure p_1 = the force on B due to the internal pressure p_2 inside the bubble. The force on B due to the pressure p_1 is given by $\pi r^2 \times p_1$, since πr^2 is the area of the circular face of B and pressure is 'force per unit area'; the force on B due to the pressure p_2 is given similarly by $\pi r^2 \times p_2$. The surface tension force acts round the *circumference* of the bubble, which has a length $2\pi r$; thus the force is $2\pi r \gamma$. It follows that

$$2\pi r \gamma + \pi r^2 p_1 = \pi r^2 p_2.$$

Simplifying, $\qquad\qquad \therefore 2\gamma = r(p_2 - p_1),$

or $\qquad\qquad p_2 - p_1 = \dfrac{2\gamma}{r}.$

Now $(p_2 - p_1)$ is the excess pressure, p, in the bubble over the outside pressure.

$$\therefore \text{excess pressure}, p, = \frac{2\gamma}{r} \qquad\qquad\qquad (1)$$

Although we considered a bubble, the same formula for the excess pressure holds for any curved liquid surface or meniscus, where r is its radius of curvature and γ is its surface tension, provided the angle of contact is zero. If the angle of contact is θ, the formula is modified by replacing γ by $\gamma \cos \theta$. Thus, in general,

$$\text{excess pressure}, p, = \frac{2\gamma \cos \theta}{r} \qquad\qquad\qquad (2)$$

Excess Pressure in Soap Bubble

A soap bubble has two liquid surfaces in contact with air, one inside the bubble and the other outside the bubble. The force on one half, B, of the bubble due to surface tension forces is thus $\gamma \times 2\pi r \times 2$, i.e., $\gamma \times 4\pi r$, Fig. 5.16. For the equilibrium of B, it follows that

$$4\pi r \gamma + \pi r^2 p_1 = \pi r^2 p_2,$$

where p_2, p_1 are the pressures inside and outside the bubble respectively. Simplifying,

$$\therefore p_2 - p_1 = \frac{4\gamma}{r},$$

$$\therefore excess\ pressure\ p = \frac{4\gamma}{r} \quad . \quad . \quad . \quad . \quad (3)$$

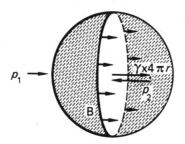

FIG. 5.16   Bubble in water

This result for excess pressure should be compared with the result obtained for a bubble formed inside a liquid, equation (1).

If γ for a soap solution is 25×10^{-3} N m^{-1}, the excess pressure inside a bubble of radius 0·5 cm or $0·5 \times 10^{-2}$ m is hence given by:

$$p = \frac{4 \times 25 \times 10^{-3}}{0·5 \times 10^{-2}} = 20\ \text{N m}^{-2}.$$

Two soap-bubbles of unequal size can be blown on the ends of a tube, communication between them being prevented by a closed tap in the middle. If the tap is opened, the *smaller* bubble is observed to collapse gradually and the size of the larger bubble increases. This can be explained from our formula $p = 4\gamma/r$, which shows that the pressure of air inside the smaller bubble is greater than that inside the larger bubble. Consequently air flows from the smaller to the larger bubble when communication is made between the bubbles, and the smaller bubble thus gradually collapses.

Since the excess pressure in a bubble is inversely-proportional to the radius, the pressure needed to form a very small bubble is high. This explains why one needs to blow hard to start a balloon growing. Once the balloon has grown, less air pressure is needed to make it expand more.

Surface Tension of Soap-Bubble

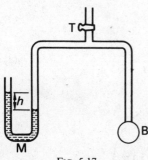

FIG. 5.17
Surface tension of soap-bubble

The surface tension of a soap solution can be measured by blowing a small soap-bubble at the end B of a tube connected to a manometer M, Fig. 5.17. The tap T is then closed, the diameter d of the bubble is measured by a travelling microscope, and the difference in levels h of the liquid in the manometer is observed with the same instrument. The excess pressure, p, in the bubble $= h\rho g$, where ρ is the density of the liquid in M.

$$\therefore h\rho g = \frac{4\gamma}{r} = \frac{4\gamma}{d/2}.$$

$$\therefore \gamma = \frac{h\rho g d}{8}.$$

Rise or Fall of Liquids in Capillary Tubes

From our knowledge of the angle of contact and the excess pressure on one side of a curved liquid surface, we can deduce that some liquids will rise in a capillary tube, whereas others will be depressed.

Suppose the tube A is placed in water, for example, Fig. 5.18 (i). At first the liquid surface becomes concave upwards in the tube, because the angle of contact with the glass is zero. Consequently the pressure on the air side, X, of the curved surface is greater than the pressure on the liquid side Y by $2\gamma/r$, where γ is the surface tension and r is the radius of curvature of the tube. But the pressure at X is atmospheric, H. Hence the pressure at Y must be less than atmospheric by $2\gamma/r$. Fig. 5.18 (i) is therefore impossible because it shows the pressure at Y equal to the atmospheric pressure. Thus, as shown in Fig. 5.18 (ii), the liquid ascends the tube to a height h such that the pressure at N is less than at M by $2\gamma/r$, Fig. 5.18 (ii). A similar argument shows that a liquid rises in a capillary tube when the angle of contact is acute. See also p. 173.

The angle of contact between mercury and glass is obtuse (p. 153). Thus when a capillary tube is placed in mercury the liquid first curves downwards. The pressure inside the liquid just below the curved surface is now greater than the pressure on the other side, which is atmospheric, and the mercury therefore moves down the tube until the excess pressure $= 2\gamma \cos\theta/r$, with the usual notation. A liquid thus falls in a capillary tube if the angle of contact is obtuse. See also p. 173.

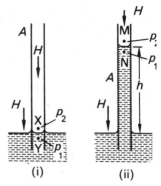

FIG. 5.18
Capillary rise by excess pressure

Capillary Rise and Fall by Pressure Method

We shall now calculate the capillary rise of water by the excess pressure formula $p = 2\gamma/r$, or $p = 2\gamma \cos\theta/r$.

In the case of a capillary tube dipping into water, the angle of contact is practically zero, Fig. 5.19 (i). Thus if p_2 is the pressure of the atmosphere, and p_1 is the pressure in the liquid, we have

$$p_2 - p_1 = \frac{2\gamma}{r}.$$

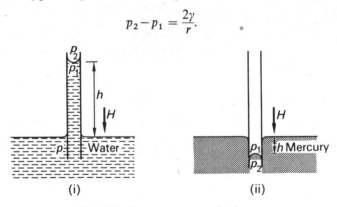

FIG. 5.19   Excess pressure application

Now if H is the atmospheric pressure, h is the height of the liquid in the tube and ρ its density,

$$p_2 = H \text{ and } p_1 = H - h\rho g,$$

$$\therefore H - (H - h\rho g) = \frac{2\gamma}{r},$$

$$\therefore h\rho g = \frac{2\gamma}{r},$$

$$\therefore h = \frac{2\gamma}{r\rho g} \qquad . \qquad . \qquad . \qquad \text{(i)}$$

The formula shows that h increases as r decreases, i.e., the narrower the tube, the greater is the height to which the water rises (see Fig. 5.7 (i), p. 152).

If the height l of the tube above the water is *less* than the calculated value of h in the above formula, the water surface at the top of the tube now meets it at an *acute angle of contact* θ. The radius of the meniscus is therefore $r/\cos\theta$, and $l\rho g = 2\gamma/(r/\cos\theta)$, or

$$l = \frac{2\gamma \cos\theta}{r\rho g} \qquad . \qquad . \qquad . \qquad \text{(ii)}$$

Dividing (ii) by (i), it follows that

$$\cos\theta = \frac{l}{h}.$$

Thus suppose water rises to a height of 10 cm in a capillary tube when it is placed in a beaker of water. If the tube is pushed down until the top is only 5 cm above the outside water surface, then $\cos\theta = \frac{5}{10} = 0.5$. Thus $\theta = 60°$. The meniscus now makes an angle of contact of $60°$ with the glass. As the tube is pushed down further, the angle of contact increases beyond $60°$. When the top of the tube is level with the water in the beaker, the meniscus in the tube becames plane. (See Example 2, p. 163.)

With Mercury in Glass

Suppose that the depression of the mercury inside a tube of radius r is h, Fig. 5.19 (ii). The pressure p_2 below the curved surface of the mercury is then greater than the (atmospheric) pressure p_1 outside the curved surface; and, from our general result,

$$p_2 - p_1 = \frac{2\gamma \cos\theta}{r},$$

where θ is the supplement of the obtuse angle of contact of mercury with glass, that is, θ is an acute angle and its cosine is positive. But $p_1 = H$ and $p_2 = H + h\rho g$, where H is the atmospheric pressure.

$$\therefore (H + h\rho g) - H = \frac{2\gamma \cos\theta}{r}.$$

SURFACE TENSION

$$\therefore hpg = \frac{2\gamma \cos \theta}{r}.$$

$$\therefore h = \frac{2\gamma \cos \theta}{r\rho g} \qquad . \qquad . \qquad . \qquad (1)$$

The height of depression, h, thus increases as the radius r of the tube decreases. See Fig. 5.7 (ii), p. 152.

EXAMPLES

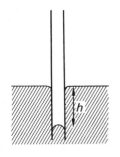

Fig. 5.20 Example

1. Define surface tension of a liquid and describe a method of finding this quantity for alcohol.

If water rises in a capillary tube 5·8 cm above the free surface of the outer liquid, what will happen to the mercury level in the same tube when it is placed in a dish of mercury? Illustrate this by the aid of a diagram. Calculate the difference in level between the mercury surfaces inside the tube and outside. (S.T. of water = 75×10^{-3} N m^{-1}. S.T. of mercury = 547×10^{-3} N m^{-1}. Angle of contact of mercury with clean glass = 130°. Density of mercury = 13600 kg m^{-3}.) (L.)

Second part. The mercury is depressed a distance h below the outside level, and is convex upward, Fig. 5.20. Suppose r is the capillary tube radius.

For water, $h = 5·8$ cm $= 5·8 \times 10^{-2}$ m, $\gamma = 75 \times 10^{-3}$ N m^{-1}, $\rho = 1000$ kg m^{-3}, $g = 9·8$ m s^{-2}.

From $\gamma = rh\rho g/2$,

$$\therefore 75 \times 10^{-3} = r \times 5·8 \times 10^{-2} \times 1000 \times 9·8/2 \text{ (r in metre)}.$$

For mercury, $\rho = 13·6 \times 10^3$ kg m^{-3}, $\gamma = 547 \times 10^{-3}$ N m^{-1}.

$$\therefore h = \frac{2\gamma \cos 50°}{r\rho g}$$

$$= \frac{2 \times 547 \times 10^{-3} \cos 50° \times 5·8 \times 10^{-2} \times 1000 \times 9·8}{13·6 \times 10^3 \times 9·8 \times 75 \times 10^{-3} \times 2}$$

$$= 0·02 \text{ m} = 2 \text{ cm}.$$

2. On what grounds would you anticipate some connection between the surface tension of a liquid and its latent heat of vaporization?

A vertical capillary tube 10 cm long tapers uniformly from an internal diameter of 1 mm at the lower end to 0·5 mm at the upper end. The lower end is just touching the surface of a pool of liquid of surface tension 6×10^{-2} N m^{-1}, density 1200 kg m^{-3} and zero angle of contact with the tube. Calculate the capillary rise, justifying your method. Explain what will happen to the meniscus if the tube is slowly lowered vertically until the upper end is level with the surface of the pool. (O. & C.)

Suppose S is the meniscus at a height h cm above the liquid surface. The tube tapers uniformly and the change in radius for a height of 10 cm is (0·05−0·025) or 0·025 cm, so that the change in radius per cm height is 0·0025 cm. Thus at a height h cm, radius of meniscus S is given by

$$r = (0·05 - 0·0025\,h) \times 10^{-2} \text{ m}$$

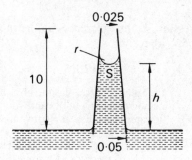

FIG. 5.21 Example

The pressure above S is atmospheric, A. The pressure below S is $(A - h\rho g)$.

$$\therefore \text{ pressure difference} = (h \times 10^{-2})\rho g = \frac{2\gamma}{r} = \frac{200\gamma}{0·05 - 0·0025h}.$$

$$\therefore 0·05h - 0·0025h^2 = \frac{200\gamma}{\rho g} = \frac{200 \times 6 \times 10^{-2}}{10^{-2} \times 1200 \times 9·8} = 0·102.$$

$$\therefore h^2 - 20h = -40 \text{ (approx.)}.$$

$$\therefore (h - 10)^2 = 100 - 40 = 60.$$

$$\therefore h = 10 - \sqrt{60} = 2·2 \text{ cm}.$$

If the tube is slowly lowered the meniscus reaches the top at some stage. On further lowering the tube the angle of contact changes from zero to an acute angle. When the upper end is level with liquid surface the meniscus becomes plane.

3. 'The surface tension of water is $7·5 \times 10^{-2}$ N m^{-1} and the angle of contact of water with glass is zero.' Explain what these statements mean. Describe an experiment to determine *either* (a) the surface tension of water, *or* (b) the angle of contact between paraffin wax and water.

A glass U-tube is inverted with the open ends of the straight limbs, of diameters respectively 0·500 mm and 1·00 mm, below the surface of water in a beaker. The air pressure in the upper part is increased until the meniscus in one limb is level with the water outside. Find the height of water in the other limb. (The density of water may be taken as 1000 kg m^{-3}.) (*L*.)

Suppose p is the air pressure inside the U-table when the meniscus Q is level with the water outside and P is the other meniscus at a height h. Let A be the atmospheric pressure. Then, if r_1 is the radius at P,

$$p - (A - h\rho g) = \frac{2\gamma}{r_1} \quad . \quad . \quad . \quad . \quad \text{(i)}$$

since the pressure in the liquid below P is $(A - h\rho g)$.

SURFACE TENSION 165

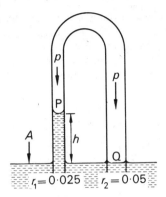

FIG. 5.22 Example (radii in cm)

The pressure in the liquid below Q = A. Hence, for Q,

$$p - A = \frac{2\gamma}{r_2} \qquad \qquad \text{(ii)}$$

where r_2 is the radius.

From (i) and (ii), it follows that

$$h\rho g = \frac{2\gamma}{r_1} - \frac{2\gamma}{r_2}$$

$$\therefore h = \frac{1}{\rho g}\left[\frac{2\gamma}{r_1} - \frac{2\gamma}{r_2}\right]$$

$$= \frac{1}{9800}\left[\frac{2 \times 0.075}{0.25 \times 10^{-3}} - \frac{2 \times 0.075}{0.5 \times 10^{-3}}\right]$$

$$= 3.1 \times 10^{-2} \text{ m (approx)}.$$

Effects of surface tension in measurements

When a hydrometer is used to measure relative density or density, the surface tension produces a downward force F on the hydrometer. If r is the radius of the stem and the angle of contact is zero, then $F = 2\pi r\gamma$. For a narrow stem, the error produced in reading the relative density from the graduations is small.

Another case of an undesirable surface tension effect occurs in measurements of the height of liquid columns in glass tubes. The height of mercury in a barometer, for example, is depressed by surface tension (p. 129). If the tubes are wide, surface tension forces can be neglected. If they are narrow, the forces must be taken into account. As an illustration, consider an inverted U-tube dipping into two liquids B and C. Fig. 5.23. These can be drawn up into the tubes to heights

h_1, h_2 respectively above the outside level. In the absence of surface tension forces, $p + h_1\rho_1 g =$ atmospheric pressure H, where p is the air pressure at the top of the tubes $= p + h_2\rho_2 g$. Thus $h_1\rho_1 = h_2\rho_2$, or $h_1/h_2 = \rho_2/\rho_1$. Thus the liquid densities may be compared from the ratio of the heights of the liquid columns.

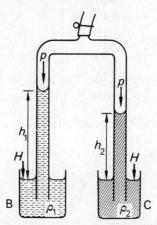

FIG. 5.23 Comparison of densities

To take account of surface tension, we proceed as follows. Using the notation on p. 161.
$$p - p_1 = \frac{2\gamma_1}{r_1},$$
where p is the air pressure at the top of the tubes, p_1 is the pressure in the liquid near the meniscus of the tube in B, γ_1 is the surface tension of the liquid, and r_1 is the radius. But, from hydrostatics, $p_1 = H - h_1\rho_1 g$.

$$\therefore p - (H - h_1\rho_1 g) = \frac{2\gamma_1}{r_1}$$

$$\therefore H - p = h_1\rho_1 g - \frac{2\gamma_1}{r_1} \qquad . \quad . \quad . \quad \text{(i)}$$

If γ_2 is the surface tension of the liquid in C, and r_2 is the radius of the tube in the liquid, then, by similar reasoning,

$$H - p = h_2\rho_2 g - \frac{2\gamma_2}{r_2} \qquad . \quad . \quad . \quad \text{(ii)}$$

From (i) and (ii),

$$\therefore h_2\rho_2 g - \frac{2\gamma_2}{r_2} = h_1\rho_1 g - \frac{2\gamma_1}{r_1}.$$

Re-arranging,

$$\therefore h_2 = \frac{\rho_1}{\rho_2} h_1 - \frac{2}{\rho_2 g}\left(\frac{\gamma_1}{r_1} - \frac{\gamma_2}{r_2}\right),$$

SURFACE TENSION   167

which is an equation of the form $y = mx+c$, where c is a constant, $h_2 = y$, $h_1 = x$, and $\rho_1/\rho_2 = m$. Thus by taking different values of h_2 and h_1, and plotting h_2 against h_1, a straight-line graph is obtained whose *slope* is equal to ρ_1/ρ_2, the ratio of the densities. In this way the effect of the surface tension can be eliminated.

Variation of Surface Tension with Temperature. Jaeger's Method

By forming a bubble inside a liquid, and measuring the excess pressure, JAEGER was able to determine the variation of the surface tension of a liquid with temperature. One form of the apparatus is shown in Fig. 5.24 (i). A capillary or drawn-out tubing A is connected to a vessel W containing a funnel C, so that air is driven slowly through A when water enters W through C. The capillary A is placed inside a beaker containing the liquid L, and a bubble forms slowly at the end of A when air is passed through it at a slow rate.

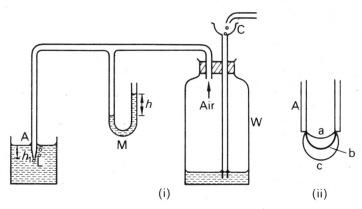

FIG. 5.24 Jaeger's method

Fig. 5.24. (ii) shows the bubble at three possible stages of growth. The radius grows from that at a to a hemispherical shape at b. Here its pressure is larger since the radius is smaller. If we consider the bubble growing to c, the radius of c would be *greater* than that of b and hence it cannot contain the increasing pressure. The downward force on the bubble due to the pressure, in fact, would be greater than the upward force due to surface tension. Hence *the bubble becomes unstable and breaks away from A when its radius is the same as that of A*. Thus as the bubble grows the pressure in it increases to a maximum, and then decreases as the bubble breaks away. The maximum pressure is observed from a manometer M containing a light oil of density ρ, and a series of observations are taken as several bubbles grow.

The maximum pressure inside the bubble $= H + h\rho g$ where h is the maximum difference in levels in the manometer M, and H is the atmospheric pressure. The pressure outside the bubble $= H + h_1\rho_1 g$, where h_1 is the depth of the orifice of A below the level of the liquid L, and ρ_1 is the latter's density.

$$\therefore \text{excess pressure} = (H + h\rho g) - (H + h_1\rho_1 g) = h\rho g - h_1\rho_1 g.$$

But
$$\text{excess pressure} = \frac{2\gamma}{r},$$

where r is the radius of the orifice of A (p. 158).

$$\therefore \frac{2\gamma}{r} = h\rho g - h_1\rho_1 g,$$

$$\therefore \gamma = \frac{rg}{2}(h\rho - h_1\rho_1).$$

By adding warm liquid to the vessel containing L, the variation of the surface tension with temperature can be determined. Experiment shows that the surface tension of liquids, and water in particular, *decreases* with increasing temperature along a fairly smooth curve. Various formulae relating the surface tension to temperature have been proposed, but none has been found to be completely satisfactory. The decrease of surface tension with temperature may be attributed to the greater average separation of the molecules at higher temperature. The force of attraction between molecules is then reduced, and hence the surface energy is reduced, as can be seen from the potential energy curve on p. 146.

Excess Pressure in Cylindrical Surface. In the most general treatment of curved liquid surfaces, it can be shown that the excess pressure p is given, for one surface, by

$$p = \gamma\left(\frac{1}{r_1} + \frac{1}{r_2}\right) \qquad \qquad (1)$$

where r_1, r_2 are the respective radii of curvature of the 'principal sections' of the surface. The principal sections are the two sections which have respectively the maximum and minimum radii of curvature. For a spherical surface such as a bubble, the principal sections are in two planes perpendicular to each other, each having a radius of curvature r. Thus $r_1 = r_2 = r$, and hence $p = 2\gamma/r$ for a bubble in water. On the other hand, a cylindrical surface has a curvature in one direction only, in a plane perpendicular to its axis; the other principal section, in the plane containing the axis, has an infinitely large radius. Thus, for a cylindrical surface,

$$p = \frac{\gamma}{r} \qquad \qquad (2)$$

The excess pressure formula for a cylindrical surface, $p = \gamma/r$, can be deduced by a method similar to that given for the spherical bubble on p. 159. Thus consider the equilibrium of one half of a cylindrical surface, obtained by drawing a plane

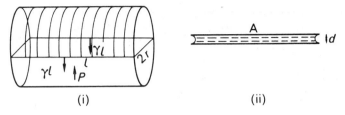

FIG. 5.25 Cylindrical surface

passing through the axis of the cylinder. Fig. 5.25 (i). Then, if l is the length of the cylinder and r the radius, the excess pressure p acts over a rectangle of area $l \times 2r$; the surface tension γ acts along sides of length l of the film. Hence, for equilibrium,

$$p \times l \times 2r = \gamma \times 2l,$$

$$\therefore p = \frac{\gamma}{r}.$$

Liquid Between Two Plates. When a small drop of water is squeezed between two plates so that a thin film of liquid is formed between them, a considerable force is required to pull the plates apart.

The magnitude of the force depends on the surface tension γ and the thickness d of the film. With a zero angle of contact, Fig. 5.25 (ii), the radius of curvature r of the film is $d/2$. Thus the atmospheric pressure is greater than the pressure inside the liquid by γ/r, or $2\gamma/d$, assuming that the radius of the liquid in contact with the plate is very large compared with d. The plates are therefore squeezed together by a force F given by

$$F = \frac{2\gamma A}{d} \quad . \quad . \quad . \quad . \quad . \quad (3)$$

where A is the area of the liquid in contact with the plates. As the thickness of the film contracts, d diminishes; the plates are hence squeezed further together, from (3). If $d = 2 \times 10^{-6}$ m, $\gamma = 7 \times 10^{-2}$ N m^{-1}, and $A = 10$ cm$^2 = 10 \times 10^{-4}$ m^2, then

$$F = \frac{2 \times 7 \times 10^{-2} \times 10 \times 10^{-4}}{2 \times 10^{-6}} = 70 \text{ N (approx)}.$$

Falling Drop. When a drop is formed at the bottom of a vertical circular tube, it can be shown that the drop becomes unstable and breaks away when the radius of the bubble is about equal to the external radius of the tube. At this stage, approximately,

upward force due to surface tension = weight of drop + downward force due to excess pressure:

$$\therefore \gamma . 2\pi r = mg + \frac{\gamma}{r} . \pi r^2,$$

as γ/r is the excess pressure in a cylindrical film. Thus if m is the mass of a drop,

$$\gamma = \frac{mg}{\pi r} \quad . \quad . \quad . \quad . \quad . \quad (4)$$

This simplified formula does not hold in practice, and Lord Rayleigh has given an approximate formula, $\gamma = mg/3.8r$, for drops formed on tubes of radii 3–5 mm. Later work showed that $mg = 2\pi r\gamma f(r/V^{1/3})$, where $f(r/V^{1/3})$ is a function of the radius r and the volume V of the drop. The weight of the drop also depends on the rate at which it is formed. The falling drop method has been used to investigate the surface tension of molten metals, as other methods are impractical.

Surface Tension and Surface Energy

We now consider the surface energy of a liquid and its relation to its surface tension γ. Consider a film of liquid stretched across a horizontal frame ABCD, Fig. 5.26. Since γ is the force per unit length, the force on the rod BC of length $l = \gamma \times 2l$, because there are two surfaces to the film.

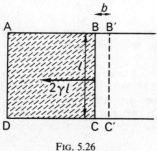

FIG. 5.26
Surface energy and work

Suppose the rod is now moved a distance b from BC to B'C' against the surface tension forces, so that the surface area of the film increases. The temperature of the film then usually decreases, in which case the surface tension alters (p. 167). If the surface area increases under *isothermal* (constant temperature) conditions, however, the surface tension is constant; and we can then say that, if γ is the surface tension at that temperature,

work done in enlarging surface area = force × distance,

$$= 2\gamma l \times b = \gamma \times 2lb.$$

But $2lb$ is the total increase in surface area of the film.

∴ work done per unit area in enlarging area = γ.

Thus the surface tension, γ, can be defined as *the work done per unit area in increasing the surface area of a liquid under isothermal conditions*. This is also called the *free surface energy*.

Surface energy and Latent heat

Inside a liquid molecules move about in all directions, continually breaking and reforming bonds with neighbours. If a molecule in the surface passes into the vapour outside, a definite amount of energy is needed to permanently break the bonds with molecules in the liquid. This amount of energy is the work done in overcoming the inward force on a molecule in the surface, discussed on p.149. Thus the energy

SURFACE TENSION 171

needed to evaporate a liquid is related to its surface energy or surface tension. The latent heat of vaporisation, which is the energy needed to change liquid to vapour at the boiling point, is therefore related to surface energy.

Surface energy

As we have seen, when the surface area of a liquid is increased, the surface energy is increased. The molecules which then reach the surface are slowed up by the inward force, so the average translational kinetic energy of all the liquid molecules is reduced. On this account the liquid cools while the surface is increased, and heat flows in from the surroundings to restore the temperature.

The increase in the *total surface energy per unit area E* is thus given by

$$E = \gamma + H \qquad . \qquad . \qquad . \qquad . \qquad . \qquad (1)$$

where H is the heat per unit area from the surroundings. Theory below shows that $H = -\theta\left(\dfrac{d\gamma}{d\theta}\right)$, where θ is the absolute temperature and $d\gamma/d\theta$ is the corresponding gradient of the γ v. θ graph, the variation of surface tension with temperature. Thus

$$E = \gamma - \theta\dfrac{d\gamma}{d\theta} \qquad . \qquad . \qquad . \qquad . \qquad (2)$$

In practice, since γ decreases with rising temperature, $d\gamma/d\theta$ is negative, and E is thus greater than γ. At 15°C, for example, $\gamma = 74 \times 10^{-3}$ N m^{-1}, $d\gamma/d\theta = -0.15 \times 10^{-3}$ N m^{-1} K^{-1}, $\theta = 288$ K. Thus, from (2),

$$E = (74 + 288 \times 0.15) \times 10^{-3} = 0.117 \text{ N m}^{-1} = 0.117 \text{ J m}^{-2}.$$

The variation of E with temperature is shown in Fig. 5.27, together with the similar variation of L, the latent heat of vaporisation (see p. 168). Both vanish at the critical temperature, since no liquid exists above the critical temperature whatever the pressure.

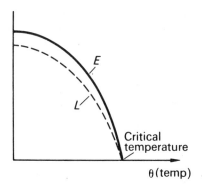

FIG. 5.27 Variation of E and L with temperature

The surface tension energy per unit area, γ, is commonly called the free *energy* because it may be changed into mechanical work and derived from mechanical work. It is the potential energy of the surface of a soap bubble, for example, when mechanical work is done in blowing the bubble. The magnitude of the free surface energy is usually much greater than the heat energy absorbed from the surroundings, and should be distinguished from the *total* surface energy, which is the sum of the free surface energy and the heat energy absorbed.

Proof of $H = -\theta\frac{d\gamma}{d\theta}$. Suppose the surface is taken round a Carnot cycle, the axes corresponding to pressure and volume for the case of a gas being replaced respectively by γ and A, where A is the surface area. Thus (1) let the area increase by unit amount isothermally at an absolute temperature θ under reversible conditions, when a quantity of heat Q_1 is absorbed and the work done is γ; (2) then let the area expand adiabatically until the temperature reaches $\theta - d\theta$, when no heat is absorbed or rejected; (3) then reduce the area isothermally by unit amount at $\theta - d\theta$ under reversible conditions, when a quantity of heat Q_2 is rejected and the work done on the film is $\gamma + d\gamma$; (4) finally, reduce the area adiabatically until the temperature θ is again reached, when the cycle is completed.

From the well-known formula for the Carnot cycle,

$$\frac{Q_1 - Q_2}{Q_1} = \frac{\theta - (\theta - d\theta)}{\theta} = \frac{d\theta}{\theta},$$

or

$$\frac{\text{net work done}}{Q_1} = \frac{d\theta}{\theta}.$$

Now $Q = H =$ heat absorbed when the area is extended isothermally by unit amount, and the net work done $= \gamma - (\gamma + d\gamma) = -d\gamma$.

$$\therefore -\frac{d\gamma}{H} = \frac{d\theta}{\theta},$$

$$\therefore H = -\theta\frac{d\gamma}{d\theta} \quad \ldots \quad \ldots \quad \text{(ii)}$$

Surface tension of solid-liquid and liquid-vapour interface

Up to the present, the surface between a liquid and vapour, called their *interface*, has been discussed. It should now be noted that a solid and a gas, or a solid and a liquid, have an interface, and each has a particular value of surface tension depending on the nature of the materials concerned.

Consider the equilibrium of a line which is normal to the plane of the paper at P, where a liquid such as water meets a solid surface such as glass at an acute angle of contact. Then, resolving forces per metre along the glass surface,

$$\gamma_{sa} = \gamma_{sl} + \gamma \cos \theta,$$

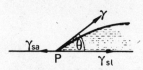

FIG. 5.28
Liquid, gas, solid inter-faces

where γ_{sa}, γ_{sl} and γ represent the surface tensions respectively between solid-air, solid-liquid and liquid-air. Thus

$$\gamma \cos \theta = \gamma_{sa} - \gamma_{sl}.$$

Note that the surface tension between solid-air, γ_{sa}, is greater than that between solid-liquid in this case. This is due to the fact that the attraction between molecules of a liquid at the interface is less than between solid and liquid. For an obtuse angle of contact, such as the case of mercury and glass, the reverse is the case. Fig. 5.28. Here the surface tension between solid-liquid is greater than between solid-air. The attraction of a mercury molecule in the interface by another mercury molecule is greater than that of the solid on the other side.

Energy applications. Rise in capillary tube

As previously seen, water spreads over or 'wets' glass. This may be explained from an energy point of view. The surface tension of a glass-water interface is less than the surface tension of a glass-air interface. Consequently the surface energy of a solid is *reduced* when the surface is covered with water and hence this is a stable system. The non-wetting of glass by mercury can be explained on a similar basis.

For the same reason, water rises up a glass capillary tube. The surface energy of the solid-air interface is reduced when it is covered with water. If the radius of the capillary tube is r and the height to which the water rises is h, then the energy released on wetting an area $2\pi rh = 2\pi rh(\gamma_{sa} - \gamma_{sl}) = 2\pi rh\gamma$, since $\gamma_{sa} - \gamma_{sl} = \gamma$ for a zero angle of contact (p. 172). Half of the energy is expended in increasing the potential energy of the liquid and the remainder in doing work against the viscous forces as the liquid rises up the tube. Thus

$$\tfrac{1}{2} \times 2\pi rh\gamma = \pi r^2 h\rho g \times \frac{h}{2}.$$

$$\therefore \gamma = \tfrac{1}{2} rh\rho g.$$

This result was derived on p. 155.

EXAMPLES

1. A soap bubble in a vacuum has a radius of 3 cm and another soap bubble in the vacuum has a radius of 6 cm. If the two bubbles coalesce under isothermal conditions, calculate the radius of the bubble formed.

Since the bubbles coalesce under isothermal conditions, the surface tension γ is constant. Suppose R is the radius in cm, $R \times 10^{-2}$ m, of the bubble formed.

Then   work done $= \gamma \times$ surface area $= \gamma \times 8\pi R^2 \times 10^{-4}$

But   original work done $= (\gamma \times 8\pi \cdot 3^2 + \gamma \times 8\pi \cdot 6^2) \times 10^{-4}$

$$\therefore \gamma \times 8\pi R^2 = \gamma \times 8\pi \cdot 3^2 + \gamma \cdot 8\pi \cdot 6^2.$$

$$\therefore R^2 = 3^2 + 6^2.$$

$$\therefore R = \sqrt{3^2 + 6^2} = 6\cdot 7 \text{ cm}.$$

2. (i) Calculate the work done against surface tension forces in blowing a soap bubble of 1 cm diameter if the surface tension of soap solution is $2\cdot 5 \times 10^{-2}$ N m^{-1}. (ii) Find the work required to break up a drop of water of radius 0·5 cm into drops of water each of radii 1 mm. (Surface tension of water $= 7 \times 10^{-2}$ N m^{-1}.)

(i) The original surface area of the bubble is zero, and the final surface area = $2 \times 4\pi r^2$ (two surfaces of bubble) = $(2 \times 4\pi \times 0.5^2) \times 10^{-4} = 2\pi \times 10^{-4}$ m^2.

$\therefore$ work done = $\gamma \times$ increase in surface area.

$$= 2.5 \times 10^{-2} \times 2\pi \times 10^{-4} = 1.57 \times 10^{-5} \text{ J}.$$

(ii) Since volume of a drop = $\frac{4}{3}\pi r^3$,

$$\text{number of drops formed} = \frac{\frac{4}{3}\pi \times 0.5^3}{\frac{4}{3}\pi \times 0.1^3} = 125.$$

$\therefore$ final total surface area of drops

$$= 125 \times 4\pi r^2 = 125 \times 4\pi \times 0.1^2 \times 10^{-4},$$
$$= 5\pi \times 10^{-4} \text{ m}^2.$$

But original surface area of drop = $4\pi \times 0.5^2 \times 10^{-4} = \pi \times 10^{-4}$ m^2.

$\therefore$ work done = $\gamma \times$ change in surface area,

$$= 7 \times 10^{-2} \times (5\pi - \pi) \times 10^{-4} = 8.8 \times 10^{-5} \text{ J}.$$

3. What is surface tension? A liquid of surface tension γ is used to form a film between a horizontal rod of length L and another shorter rod of mass M suspended from it by two light inextensible strings of equal length joining adjacent ends of each rod. The film fills the vertical plane within rods and strings.

What is the shape of each string? Show that the tension in each is $(Mg - 2\gamma L)/2 \sin \theta$, where θ is the angle which the tangent to each string makes with the upper rod. (*C.S.*)

For equilibrium of the *whole system*, the force F on the upper rod AB = Mg. Fig. 5.29 (i).

For equilibrium of the *upper rod* AB,

$$Mg = 2\gamma L \text{ (surface tension)} + 2T \sin \theta \text{ (string)}.$$

$$\therefore T = \frac{Mg - 2\gamma L}{2 \sin \theta}.$$

If a small part δs of the string is considered, the surface tension forces acts normally to it and total $2\gamma . \delta s$. Fig. 5.29 (ii). This is counterbalanced by the

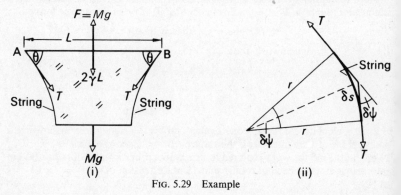

Fig. 5.29 Example

resultant force due to the tension T at either end, which is $2T\sin \delta\psi/2$, or $T.\delta\psi$ if $\delta\psi$ is small.

$$\therefore 2\gamma.\delta s = T.\delta\psi.$$

$$\therefore \frac{\delta s}{\delta\psi} = \frac{T}{2\gamma} = \text{constant}.$$

Now $r = \delta s/\delta\psi$, where r is the radius of the small part δs. Hence r is constant. Thus the string is in the shape of the *arc of a circle*.

EXERCISES 5

What are the missing words in the statements 1–8?

1. The units of surface tension are . . .
2. The dimensions of surface tension are . . .
3. Small drops of mercury are spherical because the surface area is a . . .
4. The excess pressure in a soap-bubble is given by . . .
5. The excess pressure at the meniscus of water in a capillary tube is . . .
6. A liquid will not 'wet' the surface of a solid if the angle of contact is . . .
7. Surface tension may be defined as the 'force . . .'
8. Surface tension may also be defined as the '. . . per unit area'.

Which of the following answers, A, B, C, D or E, do you consider is the correct one in the statements 9–12?

9. A molecule of a liquid which reaches the surface from the interior gains energy because *A* it reaches the surface with higher speed than when inside the liquid, *B* it overcomes a force of repulsion on molecules at the surface, *C* it overcomes a force of attraction on molecules at the surface, *D* its temperature increases, *E* the gravitational potential energy due to the earth is then higher.

10. If a section of a soap bubble through its centre is considered, the force on one half due to surface tension is *A* $2\pi r\gamma$, *B* $4\pi r\gamma$, *C* $\pi r^2\gamma$, *D* $2\gamma/r$, *E* $2\pi r^2\gamma$.

11. If water has a surface tension of 7×10^{-2} N m^{-1} and an angle of contact with water of zero, it rises in a capillary of diameter 0·5 mm to a height of *A* 70 cm, *B* 7·0 cm, *C* 6·2 cm, *D* 5·7 cm, *E* 0·5 cm.

12. In an experiment to measure the surface tension of a liquid by rise in a capillary tube which tapers, the necessary radius r would be best obtained *A* by cutting the tube at the position of the meniscus and measuring the diameter here directly, *B* by drawing up a thread of mercury of length l and using 'mass = $\pi r^2 l\rho$', *C* by measuring the diameter of the lower end of the tube with a travelling microscope, *D* by measuring the upper end of the tube with a travelling microscope, *E* by finding the average of the two measurements in C and D.

13. Define *surface tension*. A rectangular plate of dimensions 6 cm by 4 cm and thickness 2 mm is placed with its largest face flat on the surface of water. Calculate the force due to surface tension on the plate. What is the downward

force due to surface tension if the plate is placed vertical and its longest side just touches the water? (Surface tension of water = 7.0×10^{-2} N m^{-1}.)

14. What are the *dimensions* of surface tension? A capillary tube of 0.4 mm diameter is placed vertically inside (i) water of surface tension 6.5×10^{-2} N m^{-1} and zero angle of contact, (ii) a liquid of density 800 kg m^{-3}, surface tension 5.0×10^{-2} N m^{-1} and angle of contact 30°. Calculate the height to which the liquid rises in the capillary in each case.

15. Define the *angle of contact*. What do you know about the angle of contact of a liquid which (i) wets glass, (ii) does not wet glass?

A capillary tube is immersed in water of surface tension 7.0×10^{-2} N m^{-1} and rises 6.2 cm. By what depth will mercury be depressed if the same capillary is immersed in it? (Surface tension of mercury = 0.54 N m^{-1}; angle of contact between mercury and glass = 140°; density of mercury = 13 600 kg m^{-3}.)

16. (i) A soap-bubble has a diameter of 4 mm. Calculate the pressure inside it if the atmospheric pressure is 10^5 N m^{-2}. (Surface tension of soap solution = 2.8×10^{-2} N m^{-1}.) (ii) Estimate the total surface energy of a million drops of water each of radius 1.0×10^{-4} m, if the surface tension of water is 7×10^{-2} N m^{-1}. State any assumptions made.

17. Define *surface tension* of a liquid. State the units in which it is usually expressed and give its dimensions in mass, length, and time.

Derive an expression for the difference between the pressure inside and outside a spherical soap bubble. Describe a method of determining surface tension, based on the difference of pressure on the two sides of a curved liquid surface or film. (*L.*)

18. Explain briefly (*a*) the approximately spherical shape of a rain drop, (*b*) the movement of tiny particles of camphor on water, (*c*) the possibility of floating a needle on water, (*d*) why a column of water will remain in an open vertical capillary tube after the lower end has been dipped in water and withdrawn. (*N.*)

19. Define the terms surface tension, angle of contact. Describe a method for measuring the surface tension of a liquid which wets glass. List the principal sources of error and state what steps you would take to minimize them.

A glass tube whose inside diameter is 1 mm is dipped vertically into a vessel containing mercury with its lower end 1 cm below the surface. To what height will the mercury rise in the tube if the air pressure inside it is 3×10^3 N m^{-2} below atmospheric pressure? Describe the effect of allowing the pressure in the tube to increase gradually to atmospheric pressure. (Surface tension of mercury = 0.5 N m^{-1}, angle of contact with glass = 180°, density of mercury = 13600 kg m^{-3}, $g = 9.81$ m s^{-2}.) (*O. & C.*)

20. Explain how to measure the surface tension of a soap film.

The diameters of the arms of a U-tube are respectively 1 cm and 1 mm. A liquid of surface tension 7.0×10^{-2} N m^{-1} is poured into the tube which is placed vertically. Find the difference in levels in the two arms. The density may be taken as 1 000 kg m^{-3} and the contact angle zero. (*L.*)

21. Explain what is meant by surface tension, and show how its existence is accounted for by molecular theory.

Find an expression for the excess pressure inside a soap-bubble of radius R and surface tension T. Hence find the work done by the pressure in increasing the radius of the bubble from a to b. Find also the increase in surface area of the bubble, and in the light of this discuss the significance of your result. (*C*.)

22. A clean glass capillary tube, of internal diameter 0·04 cm, is held vertically with its lower end below the surface of clean water in a beaker, and with 10 cm of the tube above the surface. To what height will the water rise in the tube? What will happen if the tube is now depressed until only 5 cm of its length is above the surface? The surface tension of water is $7·2 \times 10^{-2}$ N m^{-1}.

Describe, and give the theory of some method, other than that of the rise in a capillary tube, of measuring surface tension. (*O. & C.*)

23. Explain (*a*) in terms of molecular forces why the water is drawn up above the horizontal liquid level round a steel needle which is held vertically and partly immersed in water, (*b*) why, in certain circumstances, a steel needle will rest on a water surface. In each case show the relevant forces on a diagram. (*N*.)

24. The force between two molecules may be regarded as an attractive force which increases as their separation decreases and a repulsive force which is only important at small separations and which there varies very rapidly. Draw sketch graphs (*a*) for force-separation, (*b*) for potential-energy separation. On each graph mark the equilibrium distance and on (*b*) indicate the energy which would be needed to separate two molecules initially at the equilibrium distance.

With the help of your graphs discuss briefly the resulting motion if the molecules are displaced from the equilibrium position. (*N*.)

25. Explain briefly the meaning of *surface tension* and *angle of contact*.

Account for the following: (*a*) A small needle may be placed on the surface of water in a beaker so that it 'floats', and (*b*) if a small quantity of detergent is added to the water the needle sinks.

A solid glass cylinder of length l, radius r and density σ is suspended with its axis vertical from one arm of a balance so that it is partly immersed in a liquid of density ρ. The surface tension of the liquid is γ and its angle of contact with the glass is α. If W_1 is the weight required to achieve a balance when the cylinder is in air and W_2 is the weight required to balance the cylinder when it is partly immersed with a length $h(< l)$ below the free surface of the liquid, derive an expression for the value of $W_1 - W_2$. If this method were used to measure the surface tension of a liquid, why would the result probably be less accurate than that obtained from a similar experiment using a thin glass plate? (*O. & C.*)

26. Explain in terms of molecular forces why some liquids spread over a solid surface whilst others do not.

A glass capillary tube of uniform bore of diameter 0·050 cm is held vertically with its lower end in water. Calculate the capillary rise. Describe and explain what happens if the tube is lowered so that 4·0 cm protrudes above the water surface. Assume that the surface tension of water is $7·0 \times 10^{-2}$ N m^{-1}. (*N*.)

27. Define *surface tension*. Describe how the surface tension of water at room temperature may be determined by using a capillary tube. Derive the formula used to calculate the result.

A hydrometer has a cylindrical glass stem of diameter 0·50 cm. It floats in water of density 1000 kg m^{-3} and surface tension 7.2×10^{-2} N m^{-1}. A drop of liquid detergent added to the water reduces the surface tension to 5.0×10^{-2} N m^{-1}. What will be the change in length of the exposed portion of the glass stem? Assume that the relevant angle of contact is always zero. (*N*.)

28. The lower end of a vertical clean glass capillary tube is just immersed in water. Why does water rise up the tube?

A vertical capillary tube of internal radius *r* m has its lower end dipping in water of surface tension *T* in N m^{-1}. Assuming the angle of contact between water and glass to be zero, obtain from first principles an expression for the pressure excess which must be applied to the upper end of the tube in order just to keep the water levels inside and outside the tube the same.

A capillary of internal diameter 0.7 mm is set upright in a beaker of water with one end below the surface; air is forced slowly through the tube from the upper end, which is also connected to a U-tube manometer containing a liquid of density 800 kg m^{-3}. The difference in levels on the manometer is found to build up to 9.1 cm, drop to 4.0 cm, build up to 9.1 cm again, and so on. Estimate (*a*) the depth of the open end of the capillary below the free surface of the water in the beaker, (*b*) the surface tension of water. [State clearly any assumptions you have made in arriving at these estimates.] (*O*.)

29. It is sometimes stated that, in virtue of its surface tension, the surface of a liquid behaves as if it were a stretched rubber membrane. To what extent do you think this analogy is justified?

Explain why the pressure inside a spherical soap bubble is greater than that outside. How would you investigate experimentally the relation between the excess pressure and the radius of the bubble? Show on a sketch graph the form of the variation you would expect to obtain.

If olive oil is sprayed on to the surface of a beaker of hot water, it remains as separated droplets on the water surface; as the water cools, the oil forms a continuous thin film on the surface. Suggest a reason for this phenomenon. (*C*.)

30. Describe the capillary tube method of measuring the surface tension of a liquid.

An inverted U-tube (Hare's apparatus) for measuring the specific gravity of a liquid was constructed of glass tubing of internal diameter about 2 mm. The following observations of the heights of balanced columns of water and another liquid were obtained:

Height of water (cm)	2.8	4.2	5.4	6.9	8.5	9.8	11.6
Height of liquid (cm)	2.0	3.8	5.3	7.0	9.1	10.7	13.0

Plot the above results, explain why the graph does not pass through the origin, and deduce from the graph an accurate value for the specific gravity of the liquid. (*N*.)

31. How does simple molecular theory account for surface tension? Illustrate your account by explaining the rise of water up a glass capillary.

A light wire frame in the form of a square of side 5 cm hangs vertically in water with one side in the water-surface. What additional force is necessary to pull the frame clear of the water? Explain why, if the experiment is performed with soap-solution, as the force is increased a vertical film is formed, whereas with pure water no such effect occurs. (Surface tension of water is 7.4×10^{-2} N m^{-1}.) *(O. & C.)*

32. Define *surface tension* and state the effect on the surface tension of water of raising its temperature.

Describe an experiment to measure the surface tension of water over the range of temperatures from 20°C to 70°C. Why is the usual capillary rise method unsuitable for this purpose?

Two unequal soap bubbles are formed one on each end of a tube closed in the middle by a tap. State and explain what happens when the tap is opened to put the two bubbles into connection. Give a diagram showing the bubbles when equilibrium has been reached. *(L.)*

33. Surface tension may be defined in terms of force per unit length or of energy per unit area. Show, by considering an increase in surface area of a liquid, that these definitions are equivalent. State any necessary condition.

Derive the equation $\gamma = rh\rho g/2$, where γ is the surface tension of a liquid which wets glass, h is its observed rise in a capillary tube of radius r and ρ is its density. Describe the experiment to determine γ for a liquid which is based on this formula. *(L.)*

Chapter 6

ELASTICITY

Elasticity

A bridge, when used by traffic during the day, is subjected to loads of varying magnitude. Before a steel bridge is erected, therefore, samples of the steel are sent to a research laboratory, where they undergo tests to find out whether the steel can withstand the loads to which it is likely to be subjected.

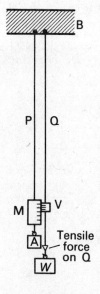

FIG. 6.1
Tensile force

Fig. 6.1 illustrates a simple laboratory method of discovering useful information about the property of steel we are discussing. Two long thin steel wires, P, Q, are suspended beside each other from a rigid support B, such as a girder at the top of the ceiling. The wire P is kept taut by a weight A attached to its end and carries a scale M graduated in centimetres. The wire Q carries a vernier scale V which is alongside the scale M.

When a load W such as 1 kgf is attached to the end of Q, the wire increases in length by an amount which can be read from the change in the reading on the vernier V. If the load is taken off and the reading on V returns to its original value, the wire is said to be **elastic** for loads from zero to 1 kgf, a term adopted by analogy with an elastic thread. When the load W is increased to 2 kgf the extension (increase in length) is obtained from V again; and if the reading on V returns to origin value when the load is removed the wire is said to be elastic at least for loads from zero to 2 kgf.

The extension of a thin wire such as Q for increasing loads may be found by experiments to be as follows:

W (kgf)	0	1	2	3	4	5	6	7	8
Extension (mm.)	0	0·14	0·28	0·42	0·56	0·70	0·85	1·01	1·19

ELASTICITY

Proportional and Elastic Limits

When the extension, e, is plotted against the load, W, a graph is obtained which is a *straight line* OA, followed by a curve ABY rising slowly at first and then very sharply, Fig. 6.2 (i). Up to A, about 5 kgf, the results show that the extension increased by 0·14 mm per kgf added to the wire. A, then, is the *proportional limit*. Along OA, and up to L just

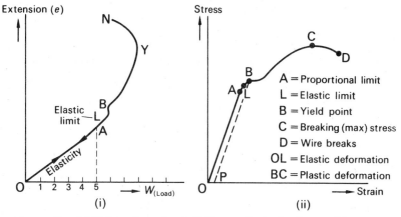

FIG. 6.2 (i) Extension v. Load. (ii) Stress v. Strain, ductile material.

beyond A, the wire returned to its original length when the load was removed. L is the *elastic limit*. Along OL the wire is said to undergo *elastic deformation*. Beyond L, however, the wire has a permanent strain OP when the stress is removed. Fig. 6.2 (ii). The reader should distinguish between the proportional limit A and the elastic limit L.

Hooke's Law

From the straight line graph OA, we deduce that *the extension is proportional to the load or tension in a wire if the proportional limit is not exceeded*. This is known as *Hooke's law*, after ROBERT HOOKE, founder of the Royal Society, who discovered the relation in 1676. The law shows that when a molecule of a solid is slightly displaced from its mean position, the restoring force is proportional to its displacement (see p. 126). One may therefore conclude that the molecules of a solid are undergoing simple harmonic motion (p. 44).

The measurements also show that it would be dangerous to load the wire with weights greater than the magnitude of the elastic limit, because the wire then suffers a permanent strain. Similar experiments in the research laboratory enable scientists to find the maximum load which a steel bridge, for example, should carry for safety. Rubber

Yield Point. Ductile and Brittle Substances. Breaking Stress

Careful experiments show that, for mild steel and iron for example, the molecules of the wire begin to 'slide' across each other soon after the load exceeds the elastic limit, that is, the material becomes *plastic*. This is indicated by the slight 'kink' at B beyond L in Fig. 6.2 (i), and it is called the *yield point* of the wire. The change from an elastic to a plastic stage is shown by a sudden increase in the extension, and as the load is increased further the extension increases rapidly along the curve YN and the wire then snaps. The *breaking stress* of the wire is the corresponding force per unit area of cross-section of the wire. Substances such as those just described, which elongate considerably and undergo plastic deformation until they break, are known as *ductile* substances. Lead, copper and wrought iron are ductile. Other substances, however, break just after the elastic limit is reached; they are known as *brittle* substances. Glass and high carbon steels are brittle.

Brass, bronze, and many alloys appear to have no yield point. These materials increase in length beyond the elastic limit as the load is increased without the appearance of a plastic stage.

The strength and ductility of a metal, its ability to flow, are dependent on defects in the metal crystal lattice. Such defects may consist of a missing atom at a site or a *dislocation* at a plane of atoms. Plastic deformation is the result of the 'slip' of atomic planes. The latter is due to the movement of dislocations, which spreads across the crystal.

Tensile Stress and Tensile Strain. Young's Modulus

We have now to consider the technical terms used in the subject of elasticity of wires. When a force or tension F is applied to the end of a wire of cross-sectional area A, Fig. 6.3,

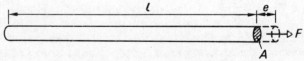

FIG. 6.3 Tensile stress and tensile strain

$$\text{the } \textit{tensile stress} = \textit{force per unit area} = \frac{F}{A} \quad . \quad (1)$$

If the extension of the wire is e, and its original length is l,

$$\text{the } \textit{tensile strain} = \textit{extension per unit length} = \frac{e}{l} \quad . \quad (2)$$

Suppose 2 kg is attached to the end of a wire of length 2 metres of diameter 0·64 mm, and the extension is 0·60 mm. Then

$$F = 2 \times 9\cdot8 \text{ N}, A = \pi \times 0\cdot032^2 \text{ cm}^2 = \pi \times 0\cdot032^2 \times 10^{-4} \text{ m}^2.$$

$$\therefore \text{ tensile stress} = \frac{2 \times 9\cdot8}{\pi \times 0\cdot032^2 \times 10^{-4}} \text{ N m}^{-2},$$

and $\quad\quad$ tensile strain $= \dfrac{0\cdot6 \times 10^{-3} \text{ metre}}{2 \text{ metre}} = 0\cdot3 \times 10^{-3}.$

It will be noted that 'stress' has units such as 'N m^{-2}'; 'strain' has no units because it is the ratio of two lengths.

A *modulus of elasticity* of the wire, called **Young's modulus (E)**, is defined as the ratio

$$E = \frac{\text{tensile stress}}{\text{tensile strain}} \quad . \quad . \quad . \quad (3)$$

Thus $\quad\quad\quad\quad\quad E = \dfrac{F/A}{e/l}.$

Using the above figures,

$$E = \frac{2 \times 9\cdot8/(\pi \times 0\cdot032^2 \times 10^{-4})}{0\cdot3 \times 10^{-3}},$$

$$= \frac{2 \times 9\cdot8}{\pi \times 0\cdot032^2 \times 10^{-4} \times 0\cdot3 \times 10^{-3}},$$

$$= 2\cdot0 \times 10^{11} \text{ N m}^{-2}.$$

It should be noted that Young's modulus, E, is calculated from the ratio stress:strain only when the wire is under 'elastic' conditions, that is, the load does not then exceed the elastic limit (p. 181). Fig. 6.2 (ii) shows the general stress-strain diagram for a ductile material.

Dimensions of Young's Modulus

As stated before, the 'strain' of a wire has no dimensions of mass, length, or time, since, by definition, it is the ratio of two lengths. Now

$$\text{dimensions of stress} = \frac{\text{dimensions of force}}{\text{dimensions of area}}$$

$$= \frac{\text{MLT}^{-2}}{\text{L}^2}$$

$$= \text{ML}^{-1}\text{T}^{-2}.$$

∴ dimensions of Young's modulus, E,

$$= \frac{\text{dimensions of stress}}{\text{dimensions of strain}}$$

$$= ML^{-1}T^{-2}.$$

Determination of Young's Modulus

The magnitude of Young's modulus for a material in the form of a wire can be found with the apparatus illustrated in Fig. 6.1, p. 180, to which the reader should now refer. The following practical points should be specially noted:

(1) The wire is made *thin* so that a moderate load of several kilogrames produces a large tensile stress. The wire is also made *long* so that a measurable extension is produced.

(2) The use of two wires, P, Q, of the same material and length, eliminates the correction for (i) the yielding of the support when loads are added to Q, (ii) changes of temperature.

(3) Both wires should be free of kinks, otherwise the increase in length cannot be accurately measured. The wires are straightened by attaching weights to their ends, as shown in Fig. 6.1.

(4) A vernier scale is necessary to measure the extension of the wire since this is always small. The 'original length' of the wire is measured from the top B *to the vernier V* by a ruler, since an error of 1 millimetre is negligible compared with an original length of several metres. For very accurate work, the extension can be measured by using a spirit level between the two wires, and adjusting a vernier screw to restore the spirit level to its original reading after a load is added.

(5) The diameter of the wire must be found by a micrometer screw gauge at several places, and the average value then calculated. The area of cross-section, $A, = \pi r^2$, where r is the radius.

(6) The readings on the vernier are also taken when the load is gradually removed in steps of 1 kilogramme; they should be very nearly the same as the readings on the vernier when the weights were added, showing that the elastic limit was not exceeded. Suppose the reading on V for loads, W, of 1 to 6 kilogramme are a, b, c, d, e, f, as follows:

W (kgf)	1	2	3	4	5	6
Reading on V	a	b	c	d	e	f

The average extension for 3 kilogramme is found by taking the average of $(d-a)$, $(e-b)$, and $(f-c)$. Young's modulus can then be calculated from the relation stress/strain, where the stress $= 3 \times 9 \cdot 8/\pi r^2$, and the strain = average extension/original length of wire (p. 182).

Magnitude of Young's Modulus

Mild steel (0·2% carbon) has a Young's modulus value of about $2·0 \times 10^{11}$ N m^{-2}, copper has a value about $1·2 \times 10^{11}$ N m^{-2}; and brass a value about $1·0 \times 10^{11}$ N m^{-2}.

The breaking stress (tenacity) of cast-iron metal is about $1·5 \times 10^8$ N m^{-2}; the breaking stress of mild steel metal is about $4·5 \times 10^8$ N m^{-2}.

At Royal Ordnance and other Ministry of Supply factories, tensile testing is carried out by placing a sample of the material in a machine known as an *extensometer*, which applies stresses of increasing value along the length of the sample and automatically measures the slight increase in length. When the elastic limit is reached, the pointer on the dial of the machine flickers, and soon after the yield point is reached the sample becomes thin at some point and then breaks. A graph showing the load v. extension is recorded automatically by a moving pen while the sample is undergoing test.

EXAMPLE

Find the maximum load in kg which may be placed on a steel wire of diameter 0·10 cm if the permitted strain must not exceed $\frac{1}{1000}$ and Young's modulus for steel is $2·0 \times 10^{11}$ N m^{-2}.

We have $\dfrac{\text{max. stress}}{\text{max. strain}} = 2 \times 10^{11}$.

$\therefore$ max. stress $= \frac{1}{1000} \times 2 \times 10^{11} = 2 \times 10^8$ N m^{-2}.

Now area of cross-section in m$^2 = \dfrac{\pi d^2}{4} = \dfrac{\pi \times 0·1^2 \times 10^{-4}}{4}$

and $\qquad\qquad\qquad$ stress $= \dfrac{\text{load } F}{\text{area}}$

$\therefore F = $ stress $\times$ area $= 2 \times 10^8 \times \dfrac{\pi \times 0·1^2 \times 10^{-4}}{4}$ N

$\qquad = 157$ N.

If $g = 10$ N/kg, load in kg $= 157/10 = 15·7$.

Force in Bar Due to Contraction or Expansion

When a bar is heated, and then prevented from contracting as it cools, a considerable force is exerted at the ends of the bar. We can derive a formula for the force if we consider a bar of Young's modulus E, a cross sectional area A, a linear expansivity of magnitude α, and a decrease in temperature of $t°C$. Then, if the original length of the bar is l, the decrease in length e if the bar were free to contract $= \alpha l t$.

Now $$E = \frac{F/A}{e/l}.$$

$$\therefore F = \frac{EAe}{l} = \frac{EA\alpha l t}{l}.$$

$$\therefore F = EA\alpha t.$$

As an illustration, suppose a steel rod of cross-sectional area $2\cdot0$ cm^2 is heated to $100°C$, and then prevented from contracting when it is cooled to $10°C$. The linear expansivity of steel $= 12 \times 10^{-6}$ K^{-1} and Young's modulus $= 2\cdot0 \times 10^{11}$ N m^{-2}. Then

$$A = 2 \text{ cm}^2 = 2 \times 10^{-4} \text{ m}^2, t = 90°C.$$

$$\therefore F = EA\alpha t = 2 \times 10^{11} \times 2 \times 10^{-4} \times 12 \times 10^{-6} \times 90 \text{ N}$$

$$= 43\,200 \text{ N}.$$

Energy Stored in a Wire

Suppose that a wire has an original length l and is stretched by a length e when a force F is applied at one end. If the elastic limit is not exceeded, the extension is directly proportional to the applied load (p. 181). Consequently the force *in the wire* has increased in magnitude from zero to F, and hence the average force in the wire while stretching was $F/2$. Now

work done = force × distance.

$\therefore$ work = average force × extension

$$= \tfrac{1}{2}Fe \qquad . \qquad . \qquad . \qquad . \qquad (1)$$

This is the amount of energy stored in the wire. The formula $\tfrac{1}{2}Fe$ gives the energy in *joule* when F is in newton and e is in metre.

Further, since $F = EAe/l$,

$$\text{energy} = \tfrac{1}{2}EA\frac{e^2}{l}.$$

As an illustration, suppose $E = 2\cdot0 \times 10^{11}$ N m^{-2}, $A = 3 \times 10^{-2}$ cm$^2 = 3 \times 10^{-6}$ m^2, $e = 1$ mm $= 1 \times 10^{-3}$ m, $l = 4$ m. Then

$$\text{energy stored} = \tfrac{1}{2}EA\frac{e^2}{l} = \tfrac{1}{2} \times \frac{2 \times 10^{11} \times 3 \times 10^{-6} \times (1 \times 10^{-3})^2}{4} \text{ joule,}$$

$$= 0\cdot075 \text{ J}.$$

The volume of the wire $= Al$. Thus, from (1),

$$\text{energy per unit volume} = \tfrac{1}{2}\frac{Fe}{Al} = \tfrac{1}{2}\frac{F}{A} \times \frac{e}{l}.$$

But $F/A =$ stress, $e/l =$ strain,

$$\therefore \text{energy per unit volume} = \tfrac{1}{2} \text{ stress} \times \text{strain} \tag{2}$$

Graph of F v. e and energy

The energy stored in the wire when it is stretched can also be found from the graph of F v. e. Fig. 6.4. Suppose the wire extension is e_1 when a load F_1 is applied, and the extension increases to e_2 when the load increases to F_2. If F is the load between F_1 and F_2 at some stage, and Δx is the small extension which then occurs, then

$$\text{energy stored} = \text{work done} = F \cdot \Delta x.$$

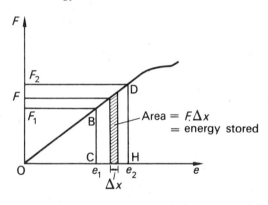

FIG. 6.4 Energy in stretched wire

Now $F.\Delta x$ is represented by the small *area* between the axis of e and the graph, shown shaded in Fig. 6.4. Thus the total work done between e_1 and e_2 is represented by the area CBDH.

If the extension occurs on the straight part of the curve, when Hooke's law is obeyed, then CBDH is a trapezium. The area of a trapezium = half the sum of the parallel sides × perpendicular distance between them $= \tfrac{1}{2}(BC + DH) \times CH = \tfrac{1}{2}(F_1 + F_2)(e_2 - e_1)$.

$\therefore$ *energy stored = average force × increase in length.*

If the extension occurs beyond the elastic limit, for example, along the curved part of the graph in Fig. 6.4, the energy expended can be obtained from the area between the curve and the axis of e.

EXAMPLES

1. A 20 kg weight is suspended from a length of copper wire 1 mm in radius. If the wire breaks suddenly, does its temperature increase or decrease? Calculate the change in temperature; Young's modulus for copper = 12×10^{10} N m^{-2}; density of copper = 9000 kg m^{-3}; specific heat capacity of copper = 420 J kg^{-1} K^{-1}.) (*C.S.*)

When the wire is stretched, it gains potential energy equal to the work done on it. When the wire is suddenly broken, this potential energy is released as the molecules return to their original position. The energy is converted into heat and thus the temperature rises.

Gain in potential energy of molecules = work done in stretching wire

$$= \tfrac{1}{2} \text{ force } (F) \times \text{extension } (e).$$

With the usual notation, $F = EA\dfrac{e}{l}$

$$\therefore e = \frac{Fl}{E.A} = \frac{(20 \times 9\cdot 8) \times l}{12 \times 10^{10} \times \pi \times (10^{-3})^2} \text{ m} = 5\cdot 2 \times 10^{-4} l \text{ m},$$

$\therefore$ potential energy gained = $\tfrac{1}{2} \times 20 \times 9\cdot 8 \times 5\cdot 2 \times 10^{-4} l = 5\cdot 1 \times 10^{-2} l$ J

Thermal capacity of wire = mass × specific heat capacity

$$= \pi \times (10^{-3})^2 \times 9000 l \times 420 = 11\cdot 9 l \text{ J K}^{-1}$$

$$\therefore \text{ temperature rise} = \frac{\text{potential energy}}{\text{thermal capacity}} = \frac{5\cdot 1 \times 10^{-2} l}{11\cdot 9 l}$$

$$= 4\cdot 3 \times 10^{-3} \text{ deg C.}$$

2. Define *stress* and *strain*. Describe the behaviour of a copper wire when it is subjected to an increasing longitudinal stress. Draw a stress-strain diagram and mark on it the elastic region, yield point and breaking stress.

A wire of length 5 m, of uniform circular cross-section of radius 1 mm is extended by 1·5 mm when subjected to a uniform tension of 100 newton. Calculate from first principles the strain energy per unit volume assuming that deformation obeys Hooke's law.

Show how the stress-strain diagram may be used to calculate the work done in producing a given strain, when the material is stretched beyond the Hooke's law region. (*O. & C.*)

$$\text{Strain energy} = \tfrac{1}{2} \text{ tension} \times \text{extension}$$

Tension = 100 newton. Extension = $1\cdot 5 \times 10^{-3}$ m.

$$\therefore \text{ energy} = \tfrac{1}{2} \times 100 \times 1\cdot 5 \times 10^{-3} = 0\cdot 075 \text{ J}.$$

Volume of wire = length × area = $5 \times \pi \times 1 \times 10^{-6}$ m^3.

$$\therefore \text{ energy per unit volume} = \frac{0\cdot 075}{5 \times \pi \times 1 \times 10^{-6}} = 4\cdot 7 \times 10^3 \text{ J m}^{-3} \text{ (approx.)}.$$

Bulk Modulus

When a gas or a liquid is subjected to an increased pressure the substance contracts. A change in bulk thus occurs, and the *bulk strain* is defined by:

$$\text{strain} = \frac{\text{change in volume}}{\text{original volume}}.$$

The *bulk stress* on the substance is the increased force per unit area, by definition, and the bulk modulus, K, is given by:

$$K = \frac{\text{bulk stress}}{\text{bulk strain}}$$

$$= \frac{\text{increase in force per unit area}}{\text{change in volume/original volume}}.$$

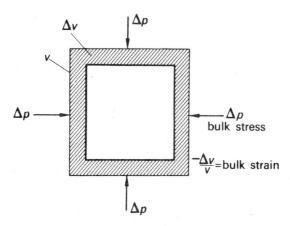

FIG. 6.5 Bulk stress and bulk strain

If the original volume of the substance is v, the change in volume may be denoted by $-\Delta v$ when the pressure increases by a small amount Δp; the minus indicates that the volume decreases. Thus (Fig. 6.5)

$$K = -\frac{\Delta p}{\Delta v/v}.$$

When δp and δv become very small, then, in the limit,

$$K = -v\frac{dp}{dv} \qquad . \qquad . \qquad . \qquad . \qquad (1)$$

The bulk modulus of water is about 2×10^9 N m^{-2} for pressures in the range $1-25$ atmospheres; the bulk modulus of mercury is about

27×10^9 N m^{-2}. The bulk modulus of gases depends on the pressure, as now explained. Generally, since the volume change is relatively large, the bulk modulus of a gas is low compared with that of a liquid.

Bulk Modulus of a Gas

If the pressure, p, and volume, v, of a gas change under conditions such that
$$pv = \text{constant},$$
which is Boyle's law, the changes are said to be *isothermal* ones. In this case, by differentiating the product pv with respect to v, we have
$$p + v\frac{dp}{dv} = 0.$$
$$\therefore p = -v\frac{dp}{dv}.$$

But the bulk modulus, K, of the gas is equal to $-v\dfrac{dp}{dv}$ by definition (see p. 189).
$$\therefore K = p \qquad . \qquad . \qquad . \qquad . \qquad (2)$$

Thus the *isothermal bulk modulus is equal to the pressure*.

When the pressure, p, and volume, v, of a gas change under conditions such that
$$pv^\gamma = \text{constant},$$
where $\gamma = c_p/c_V$ = the ratio of the specific heat capacities of the gas, the changes are said to be *adiabatic* ones. This equation is the one obeyed by local values of pressure and volume in air when a sound wave travels through it. Differentiating both sides with respect to v,
$$\therefore p \times \gamma v^{\gamma-1} + v^\gamma \frac{dp}{dv} = 0,$$
$$\therefore \gamma p = -v\frac{dp}{dv},$$
$$\therefore \text{adiabatic bulk modulus} = \gamma p \qquad . \qquad . \qquad . \qquad . \qquad (3)$$

For air at normal pressure, $K = 10^5$ N m^{-2} isothermally and 1.4×10^5 N m^{-2} adiabatically. The values of K are of the order 10^5 times smaller than liquids as gases are much more compressible.

Velocity of Sound

The velocity of sound waves through any material depends on (i) its density ρ, (ii) its modulus of elasticity, E. Thus if V is the velocity, we

ELASTICITY

may say that

$$V = kE^x\rho^y \qquad (i),$$

where k is a constant and x, y are indices we can find by the theory of dimensions (p. 34).

The units of velocity, V, are LT^{-1}; the units of density ρ are ML^{-3}; and the units of modulus of elasticity, E, are $ML^{-1}T^{-2}$ (see p. 183). Equating the dimensions on both sides of (i),

$$\therefore LT^{-1} = (ML^{-1}T^{-2})^x \times (ML^{-3})^y.$$

Equating the indices of M, L, T on both sides, we have

$$0 = x + y,$$
$$1 = -x - 3y,$$
$$-1 = -2x.$$

Solving, we find $x = \frac{1}{2}$, $y = -\frac{1}{2}$. Thus $V = kE^{\frac{1}{2}}\rho^{-\frac{1}{2}}$. A rigid investigation shows $k = 1$, and thus

$$V = E^{\frac{1}{2}}\rho^{-\frac{1}{2}} = \sqrt{\frac{E}{\rho}}.$$

In the case of a solid, E is Young's modulus. In the case of air and other gases, and of liquids, E is replaced by the bulk modulus K. Laplace showed that the adiabatic bulk modulus must be used in the case of a gas, and since this is γp, the velocity of sound in a gas is given by the expression

$$V = \sqrt{\frac{\gamma p}{\rho}}.$$

Modulus of Rigidity or Shear Modulus

So far we have considered the strain in one direction, or tensile strain, to which Young's modulus is applicable and the strain in bulk or volume, to which the bulk modulus is applicable.

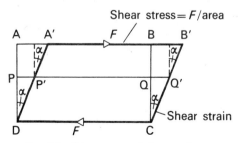

FIG. 6.6 Shear stress and shear strain

Consider a block of material ABCD, such as pitch or plastic for convenience. Fig. 6.6. Suppose the lower plane CD is fixed, and a stress parallel to CD is applied by a force F to the upper side AB. The block then changes its shape and takes up a position A'B'CD. It can now be seen that planes in the material parallel to DC are displaced relative to each other. The plane AB, for example, which was originally directly opposite the plane PQ, is displaced to A'B' and PQ is displaced to P'Q'. The *angular displacement* α is defined as the *shear strain*. α is the angular displacement between any two planes, for example, between CD and P'Q'.

No volume change occurs in Fig. 6.6. Further, since the force along CD is F in magnitude, it forms a *couple* with the force F applied to the upper side AB. The *shear stress* is defined as the 'shear force per unit area' on the face AB (or CD), as in Young's modulus or the bulk modulus. Unlike the case for these modulii, however, the shear stress has a turning or 'displacement' effect owing to the couple present. The solid does not collapse because in a strained equilibrium position such as A'B'CD in Fig. 6.6, the external couple acting on the solid due to the forces F is balanced by an opposing couple due to stresses inside the material.

If the elastic limit is not exceeded when a shear stress is applied, that is, the solid recovers its original shape when the stress is removed, the *modulus of rigidity* or *shear modulus*, G, is defined by:

$$G = \frac{\text{shear stress (force per unit area)}}{\text{shear strain (angular displacement, } \alpha)}.$$

Shear strain has no units; shear stress has units of newton m^{-2}. The modulus of rigidity of copper is 4.8×10^{10} N m^{-2}; for phosphor-bronze it is 4.4×10^{10} N m^{-2}, and for quartz fibre it is 3.0×10^{10} N m^{-2}.

If a spiral spring is stretched, all parts of the spiral become twisted. The applied force has thus developed a 'torsional' or shear strain. The extension of the spring hence depends on its modulus of rigidity, in addition to its dimensions.

Torsion wire

In sensitive current-measuring instruments, a very weak control is needed for the rotation of the instrument coil. This may be provided by using a long elastic or *torsion wire* of phosphor bronze in place of a spring. The coil is suspended from the lower end of the wire and when it rotates through an angle θ, the wire sets up a weak opposing couple equal to $c\theta$, where c is the elastic constant of the wire. Quartz fibres are very fine but comparitively strong, and have elastic properties. They are also used for sensitive control (see p. 69).

ELASTICITY

The magnitude of c, the elastic constant, can be derived as follows. Consider a wire of radius a, length l, modulus of rigidity G, fixed at the upper end and twisted by a couple of moment C at the other end. If we take a section of the cylindrical wire between radii r and $r+\delta r$, then a 'slice' of the material ODBX has been sheared through an angle α to a position ODB_1X where X is the centre of the lower end of the wire. Fig. 6.7. From the definition of modulus of rigidity, $G =$ torsional stress $\div$ torsional strain $= F/A \div \alpha$, where F is the tangential force applied over an area A.

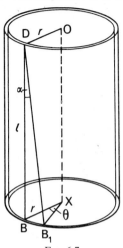

FIG. 6.7
Shear (Torsion) in wire

Now $A =$ area of circular annulus at lower end $= 2\pi r . \delta r$.

$$\therefore F = GA\alpha = G.2\pi r.\delta r.\alpha.$$

From Fig. 6.7, it follows that $BB_1 = l\alpha$, and $BB_1 = r\theta$.

$$\therefore l\alpha = r\theta, \text{ or } \alpha = r\theta/l.$$

$$\therefore F = \frac{G.2\pi r\ \delta r.r\theta}{l} = \frac{2\pi G\theta r^2.\delta r}{l}.$$

$\therefore$ moment of F about axis OX of wire $= F.r$

$$= \frac{2\pi G\theta}{l}.r^3.\delta r.$$

$\therefore$ total moment, or couple torque C,

$$= \int_0^a \frac{2\pi G\theta}{l}.r^3 dr = \frac{2\pi G\theta}{l}.\frac{a^4}{4}$$

$$\therefore C = \frac{\pi Ga^4\theta}{2l} \qquad \qquad \qquad \qquad \text{(i)}$$

If the wire is a hollow cylinder of radii a, b respectively, the limits of integration are altered accordingly, and

$$\text{moment of couple} = \int_a^b \frac{2\pi G\theta}{l}.r^3 dr = \frac{\pi G(b^4-a^4)\theta}{2l}.$$

Determinations of modulus of rigidity. Dynamical method. One method of measuring the modulus of rigidity of a wire E is to clamp it vertically at one end, attach a horizontal disc D of known moment of inertia, I, at the other end, and then time the horizontal torsional oscillations of D. Fig. 6.8(i). On p. 102, it was shown that the period of oscillation, $T, = 2\pi\sqrt{I/c}$, where c is the opposing couple per unit angle of twist. Thus, with our previous notation, as $\theta = 1$,

$$c = \frac{\pi Ga^4}{2l}.$$

$$\therefore T = 2\pi\sqrt{\frac{2lI}{\pi Ga^4}}.$$

or
$$G = \frac{8\pi l I}{a^4 T^2}$$

Hence G can be evaluated from measurements of l, a, I, T.

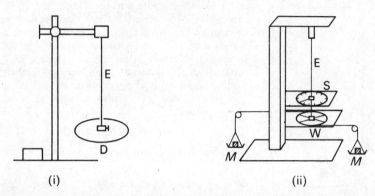

FIG. 6.8 Modulus of rigidity measurement

Statical method. The modulus of rigidity, G, of the wire E can also be found by measuring the steady deflection θ at the lower end on a scale S graduated in degrees when a couple is applied round a wheel W. Fig. 6.8 (ii). If M is the mass in each scale-pan, and d is the diameter of W, the moment of the couple on the wire $= Mgd = \pi G a^4 \theta / 2l$. The angle θ in radians, and a, l, are known, and hence G can be evaluated.

Poisson's Ratio

When a rubber cord is extended its diameter usually decreases at the same time. *Poisson's ratio*, σ, is the name given to the ratio

$$\frac{\text{lateral contraction/original diameter}}{\text{longitudinal extension/original length}} \qquad . \qquad . \qquad (1)$$

and is a constant for a given material. If the original length of a rubber strip is 100 cm and it is stretched to 102 cm, the fractional longitudinal extension = 2/100. If the original diameter of the cord is 0·5 cm and it decreases to 0·495 cm, the fractional lateral contraction = 0·005/0·5 = 1/100. Thus, from the definition of Poisson's ratio,

$$\sigma = \frac{1/100}{2/100} = \tfrac{1}{2}.$$

When the *volume* of a strip of material remains *constant* while an extension and a lateral contraction takes place, it can easily be shown that Poisson's ratio is 0·5 in this case. Thus suppose that the length of the strip is l and the radius is r.

Then
volume, V, $= \pi r^2 l$.

By differentiating both sides, noting that V is a constant and that we have a product of variables on the right side,

$$\therefore 0 = \pi r^2 \times \delta l + l \times 2\pi r \delta r.$$

$$\therefore r\delta l = -2l\delta r.$$

$$\therefore -\frac{\delta r/r}{\delta l/l} = \tfrac{1}{2}.$$

But $-\delta r/r$ is the lateral contraction in radius/original radius, and $\delta l/l$ is the longitudinal extension/original length.

$$\therefore \text{Poisson's ratio, } \sigma, = \tfrac{1}{2}.$$

Experiments show that σ is 0·48 for rubber, 0·29 for steel, 0·27 for iron, and 0·26 for copper. Thus the three metals increase in volume when stretched, whereas rubber remains almost unchanged in volume.

Summary

The three modulii of elasticity are compared in the table below:

	Young's modulus, E	Modulus of Rigidity, G	Bulk modulus, K
1.	Definition: $\dfrac{\text{tensile stress}}{\text{tensile strain}}$	$\dfrac{\text{shear stress}}{\text{shear strain}}$	Definition: $\dfrac{\text{pressure change}}{-\Delta v/v}$
2.	Relates to change in *length* ('tensile')	Relates to change in *shape* ('shear')	Relates to change in *volume* ('bulk')
3.	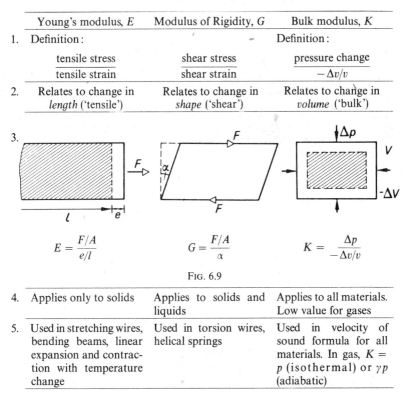		
4.	Applies only to solids	Applies to solids and liquids	Applies to all materials. Low value for gases
5.	Used in stretching wires, bending beams, linear expansion and contraction with temperature change	Used in torsion wires, helical springs	Used in velocity of sound formula for all materials. In gas, $K = p$ (isothermal) or γp (adiabatic)

FIG. 6.9

EXERCISES 6

(*Assume g* = 9·8 m s^{-2} *unless otherwise stated*)

What are the missing words in the statements 1–6?

1. When a weight is attached to a suspended long wire, it produces a ... strain.
2. The units of Young's modulus are ...
3. In measuring Young's modulus, the ... must not be exceeded.
4. The energy gained by a wire when stretched = ... × extension.
5. Bulk stress is defined as the ... change.
6. When a wire is twisted, a ... strain is produced.

Which of the following answers, A, B, C, D or E, do you consider is the correct one in the statements 7–10?

7. If a metal bar, coefficient of linear expansion α, Young's modulus E, area of cross-section A and length l, is heated through t°C when clamped at both ends, the force in the bar is calculated from A $EAlt$, B EAt/α, C $EA\alpha t$, D $E^2 A\alpha l$, E $A\alpha tl$.

8. When a spiral spring is stretched by a weight attached to it, the strain is A tensile, B shear, C bulk, D elastic, E plastic.

9. The energy in a stretched wire is A $\frac{1}{2}$ load × extension, B load × extension, C stress × extension, D load × strain, E $\frac{1}{2}$ load × strain.

10. In an experiment to measure Young's modulus, the wire is thin and long so that A very heavy weights can be attached, B the wire can be suspended from the ceiling, C another identical wire can be arranged parallel to it, D the stress is large and the extension is measurable for laboratory loads, E a micrometer gauge can be used for accurate measurement.

11. Define *tensile stress, tensile strain, Young's modulus*. What are the units and dimensions of each?

A load of 2 kg is applied to the ends of a wire 4 m long, and produces an extension of 0·24 mm. If the diameter of the wire is 2 mm, calculate the stress on the wire, its strain, and the value of Young's modulus.

12. What load in kilogramme must be applied to a steel wire 6 m long and diameter 1·6 mm to produce an extension of 1 mm? (Young's modulus for steel = $2·0 \times 10^{11}$ N m^{-2}.)

13. Find the extension produced in a copper wire of length 2 m and diameter 3 mm when a load of 3 kg is applied. (Young's modulus for copper = $1·1 \times 10^{11}$ N m^{-2}.)

14. What is meant by (i) elastic limit, (ii) Hooke's law, (iii) yield point, (iv) perfectly elastic? Draw sketches of stress *v.* strain to illustrate your answers.

15. 'In an experiment to determine Young's modulus, the strain should not

exceed 1 *in* 1000.' Explain why this limitation is necessary and describe an experiment to determine Young's modulus for the material of a metal wire.

In such an experiment, a brass wire of diameter 0·0950 cm is used. If Young's modulus for brass is 9.86×10^{10} N m^{-2}, find the greatest permissible load on the wire. (*L*.)

16. Define *stress* and *strain*, and explain why these quantities are useful in studying the elastic behaviour of a material.

State one advantage and one disadvantage in using a long wire rather than a short stout bar when measuring Young's modulus by direct stretching.

Calculate the minimum tension with which platinum wire of diameter 0·1 mm must be mounted between two points in a stout invar frame if the wire is to remain taut when the temperature rises 100 K. Platinum has coefficient of linear expansion 9×10^{-6} K^{-1} and Young's modulus 17×10^{10} N m^{-2}. The thermal expansion of invar may be neglected. (*O. & C.*)

17. Explain the terms *stress, strain, modulus of elasticity* and *elastic limit*. Derive an expression in terms of the tensile force and extension for the energy stored in a stretched rubber cord which obeys Hooke's law.

The rubber cord of a catapult has a cross-sectional area 1·0 mm^2 and a total unstretched length 10·0 cm. It is stretched to 12·0 cm and then released to project a missile of mass 5·0 g. From energy considerations, or otherwise, calculate the velocity of projection, taking Young's modulus for the rubber as 5.0×10^8 N m^{-2}. State the assumptions made in your calculation.

18. State Hooke's law, and describe in detail how it may be verified experimentally for copper wire. A copper wire, 200 cm long and 1·22 mm diameter, is fixed horizontally to two rigid supports 200 cm long. Find the mass in grams of the load which, when suspended at the mid-point of the wire, produces a sag of 2 cm at that point. Young's modulus for copper = 12.3×10^{10} N m^{-2}. (*L*.)

19. Distinguish between Young's modulus, the bulk modulus and the shear modulus of a material. Describe a method for measuring Young's modulus. Discuss the probable sources of error and assess the magnitude of the contribution from each.

A piece of copper wire has twice the radius of a piece of steel wire. Young's modulus for steel is twice that for the copper. One end of the copper wire is joined to one end of the steel wire so that both can be subjected to the same longitudinal force. By what fraction of its length will the steel have stretched when the length of the copper has increased by 1%? (*O. & C.*)

20. In an experiment to measure Young's modulus for steel a wire is suspended vertically and loaded at the free end. In such an experiment, (*a*) why is the wire long and thin, (*b*) why is a second steel wire suspended adjacent to the first?

Sketch the graph you would expect to obtain in such an experiment showing the relation between the applied load and the extension of the wire. Show how it is possible to use the graph to determine (*a*) Young's modulus for the wire, (*b*) the work done in stretching the wire.

If Young's modulus for steel is 2.00×10^{11} N m^{-2}, calculate the work done in stretching a steel wire 100 cm in length and of cross-sectional area 0·030 cm^2 when a load of 10 kg is slowly applied without the elastic limit being reached. (*N*.)

21. Describe the changes which take place when a wire is subjected to a steadily increasing tension. Include in your description a sketch graph of tension against extension for (a) a ductile material such as drawn copper and (b) a brittle one such as cast iron.

Show that the energy stored in a rod of length L when it is extended by a length l is $\frac{1}{2}El^2/L^2$ per unit volume where E is Young's modulus of the material.

A railway track uses long welded steel rails which are prevented from expanding by friction in the clamps. If the cross-sectional area of each rail is 75 cm² what is the elastic energy stored per kilometre of track when its temperature is raised by 10°C? (Coefficient of thermal expansion of steel = 1.2×10^{-5} K^{-1}; Young's modulus for steel = 2×10^{11} N m^{-2}.) (*O. & C.*)

22. What is meant by saying that a substance is 'elastic'?

A vertical brass rod of circular section is loaded by placing a 5 kg weight on top of it. If its length is 50 cm, its radius of cross-section 1 cm, and the Young's modulus of the material 3.5×10^{10} N m^{-2}, find (a) the contraction of the rod, (b) the energy stored in it. (*C.*)

23. Give a short account of what happens when a copper wire is stretched under a gradually increasing load. What is meant by *modulus of elasticity, elastic limit, perfectly elastic*?

When a rubber cord is stretched the change in volume is very small compared with the change in shape. What will be the numerical value of Poisson's ratio for rubber, i.e., the ratio of the fractional decrease in diameter of the stretched cord to its fractional increase in length? (*L.*)

24. Describe an accurate method of determining Young's modulus for a wire. Detail the precautions necessary to avoid error, and estimate the accuracy attainable.

A steel tyre is heated and slipped on to a wheel of radius 40 cm which it fits exactly at a temperature t°C. What is the maximum value of t if the tyre is not to be stretched beyond its elastic limit when it has cooled to air temperature (17°C)? What will then be the tension in the tyre, assuming it to be 4 cm wide and 3 mm thick? The value of Young's modulus for steel is 1.96×10^{11} N m^{-2}, its coefficient of linear expansion is 1.1×10^{-5} K^{-1}, and its elastic limit occurs for a stress of 2.75×10^8 N m^{-2}. The wheel may be assumed to be at air temperature throughout, and to be incompressible. (*O. & C.*)

25. State Hooke's law and describe, with the help of a rough graph, the behaviour of a copper wire which hangs vertically and is loaded with a gradually increasing load until it finally breaks. Describe the effect of gradually reducing the load to zero (a) before, (b) after the elastic limit has been reached.

A uniform steel wire of density 7800 kg m^{-3} weighs 16 g and is 250 cm long. It lengthens by 1.2 mm when stretched by a load of 8 kg. Calculate (a) the value of Young's modulus for the steel, (b) the energy stored in the wire. (*N.*)

26. Describe an experimental method for the determination of (a) Young's modulus, (b) the elastic limit, of a metal in the form of a thin wire.

A steel rod of mass 97.5 g and of length 50 cm is heated to 200°C and its ends securely clamped. Calculate the tension in the rod when its temperature is reduced to 0°C, explaining how the calculation is made. (Young's modulus for steel $= 2.0 \times 10^{11}$ N m^{-2}; linear expansivity $= 1.1 \times 10^{-5}$ K^{-1}; density of steel $= 7800$ kg m^{-3}.) (*L.*)

27. What do you understand by Hooke's law of elasticity? Describe how you would verify it in any particular case.

A wire of radius 0·2 mm is extended by 0·1% of its length when it supports a load of 1 kg; calculate Young's modulus for the material of the wire. (*L.*)

28. Define Young's modulus of elasticity. Describe an accurate method of determining it. The rubber cord of a catapult is pulled back until its original length has been doubled. Assuming that the cross-section of the cord is 2 mm square, and that Young's modulus for rubber is 10^7 N m^{-2} calculate the tension in the cord. If the two arms of the catapult are 6 cm apart, and the unstretched length of the cord is 8 cm what is the stretching force? (*O. & C.*)

29. Define *Young's modulus of elasticity* and *coefficient of linear expansion*. State units in which each may be expressed and describe an experimental determination of Young's modulus.

For steel. Young's modulus is 1.8×10^{11} N m^{-2} and the coefficient of expansion 1.1×10^{-5} K^{-1}. A steel wire 1 mm in diameter is stretched between two supports when its temperature is 200°C. By how much will the force the wire exerts on the supports increase when it cools to 20°C, if they do not yield? (*L.*)

30. Define *elastic limit* and *Young's modulus* and describe how you would find the values for a copper wire.

What stress would cause a wire to increase in length by one-tenth of one per cent if Young's modulus for the wire is 12×10^{10} N m^{-2}? What load in kg would produce this stress if the diameter of the wire is 0·56 mm? (*L.*)

Chapter 7

SOLID FRICTION. VISCOSITY

SOLID FRICTION

Static Friction

WHEN a person walks along a road, he or she is prevented from slipping by the force of friction at the ground. In the absence of friction, for example on an icy surface, the person's shoe would slip when placed on the ground. The frictional force always *opposes* the motion of the shoe.

The frictional force between the surface of a table and a block of wood A can be investigated by attaching one end of a string to A and the other to a scale-pan S, Fig. 7.1. The string passes over a fixed

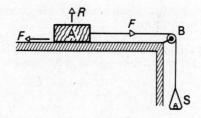

FIG. 7.1 Solid friction

grooved wheel B. When small weights are added to S, the block does not move. The frictional force between the block and table is thus equal to the total weight on S together with the weight of S. When more weights are added, A does not move, showing that the frictional force has increased, but as the weight is increased further, A suddenly begins to slip. The frictional force now present between the surfaces is called the *limiting frictional force*, and we are said to have reached *limiting friction*. The limiting frictional force is the maximum frictional force between the surfaces.

Coefficient of Static Friction

The normal reaction, R, of the table on A is equal to the weight of A. By placing various weights on A to alter the magnitude of R, we can find how the limiting frictional force F varies with R by the experiment just described. The results show that, approximately,

$$\frac{\text{limiting frictional force }(F)}{\text{normal reaction }(R)} = \mu, \text{ a constant,}$$

and μ is known as the *coefficient of static friction* between the two surfaces. The magnitude of μ depends on the nature of the two surfaces; for example it is about 0·2 to 0·5 for wood on wood, and about 0·2 to 0·6 for wood on metals. Experiment also shows that the limiting frictional force is the same if the block A in Fig. 7.1 is turned on one side so that its surface area of contact with the table decreases, and thus the limiting frictional force is independent of the area of contact when the normal reaction is the same.

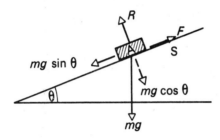

FIG. 7.2 Coefficient by inclined plane

The coefficient of static friction, μ, can also be found by placing the block A on the surface S, and then gently tilting S until A is on the point of slipping down the plane, Fig. 7.2. The static frictional force F is then equal to $mg \sin \theta$, where θ is the angle of inclination of the plane to the horizontal; the normal reaction R is equal to $mg \cos \theta$.

$$\therefore \mu = \frac{F}{R} = \frac{mg \sin \theta}{mg \cos \theta} = \tan \theta,$$

and hence μ can be found by measuring θ.

Kinetic Friction. Coefficient of Kinetic (Dynamic) Friction

When brakes are applied to a bicycle, a frictional force is exerted between the moving wheels and brake blocks. In contrast to the case of static friction, when one of the objects is just on the point of slipping, the frictional force between the moving wheel and brake blocks is called a *kinetic (or dynamic) frictional force*. Kinetic friction thus occurs between two surfaces which have relative motion.

The *coefficient of kinetic (dynamic) friction*, μ', between two surfaces is defined by the relation

$$\mu' = \frac{F'}{R},$$

where F' is the frictional force when the object moves with a uniform velocity and R is the normal reaction between the surfaces. The coefficient of kinetic friction between a block A and a table can be found by the apparatus shown in Fig. 7.1. Weights are added to the scale-pan, and each time A is given a slight push. At one stage A continues to move with a constant velocity, and the kinetic frictional force F' is then equal to the total weight in the scale-pan together with the latter's weight. On dividing F' by the weight of A, the coefficient can be calculated. Experiment shows that, when weights are placed on A to vary the normal reaction R, the magnitude of the ratio F'/R is approximately constant. Results also show that the coefficient of kinetic friction between two given surfaces is less than the coefficient of static friction between the same surfaces, and that the coefficient of kinetic friction between two given surfaces is approximately independent of their relative velocity.

Laws of Solid Friction

Experimental results on solid friction are summarised in the *laws of friction*, which state:

(1) The frictional force between two surfaces opposes their relative motion.

(2) The frictional force is independent of the area of contact of the given surfaces when the normal reaction is constant.

(3) The limiting frictional force is proportional to the normal reaction for the case of static friction. The frictional force is proportional to the normal reaction for the case of kinetic (dynamic) friction, and is independent of the relative velocity of the surfaces.

Theory of Solid Friction

The laws of solid friction were known hundreds of years ago, but they have been explained only in comparatively recent years, mainly by F. P. Bowden and collaborators. Sensitive methods, based on electrical conductivity measurements, reveal that the true area of contact between two surfaces is extremely small, perhaps one ten-thousandth of the area actually placed together for steel surfaces. This

SOLID FRICTION, VISCOSITY 203

is explained by photographs which show that some of the atoms of a metal project slightly above the surface, making a number of crests or 'humps'. As Bowden has stated: 'The finest mirror, which is flat to a millionth of a centimetre, would to anyone of atomic size look rather like the South Downs—valley and rolling hills a hundred or more atoms high.' Two metal surfaces thus rest on each others projections when placed one on the other.

Since the area of actual contact is extremely small, the pressures at the points of contact are very high, perhaps 1000 million kgf per m^2 for steel surfaces. The projections merge a little under the high pressure, producing adhesion or 'welding' at the points, and a force which opposes motion is therefore obtained. This explains Law 1 of the laws of solid friction. When one of the objects is turned over, so that a smaller or larger surface is presented to the other object, measurements show that the small area of actual contact remains constant. Thus the frictional force is independent of the area of the surfaces, which explains Law 2. When the load increases the tiny projections are further squeezed by the enormous pressures until the new area of contact becomes big enough to support the load. The greater the load, the greater is the area of actual contact, and the frictional force is thus approximately proportional to the load, which explains Law 3.

VISCOSITY

If we move through a pool of water we experience a resistance to our motion. This shows that there is a *frictional force* in liquids. We say this is due to the **viscosity** of the liquid. If the frictional force is comparatively low, as in water, the viscosity of the liquid is low; if the frictional force is large, as in glue or glycerine, the viscosity of the liquid is high. We can compare roughly the viscosity of two liquids by filling two measuring cylinders with each of them, and allowing identical small steel ball-bearings to fall through each liquid. The sphere falls more slowly through the liquid of higher viscosity.

As we shall see later, the viscosity of a lubricating oil is one of the factors which decide whether it is suitable for use in an engine. The Ministry of Aircraft Production, for example, listed viscosity values to which lubricating oils for aero-engines must conform. The subject of viscosity has thus considerable practical importance.

Newton's Formula. Coefficient of Viscosity

When water flows slowly and steadily through a pipe, the layer A of the liquid in contact with the pipe is practically stationary, but the central part C of the water is moving relatively fast, Fig. 7.3. At other layers between A and C, such as B, the water has a velocity less than at

C, the magnitude of the velocities being represented by the length of the arrowed lines in Fig. 7.3. Now as in the case of two solid surfaces

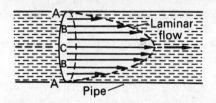

FIG. 7.3   Laminar (uniform) flow through pipe

moving over each other, a frictional force is exerted between two liquid layers when they move over each other. Thus because the velocities of neighbouring layers are different, as shown in Fig. 7.3, a frictional force occurs between the various layers of a liquid when flowing through a pipe.

The basic formula for the frictional force, F, in a liquid was first suggested by NEWTON. He saw that the larger the *area* of the surface of liquid considered, the greater was the frictional force F. He also stated that F was directly proportional to the *velocity gradient* at the part of the liquid considered. This is the case for most common liquids, called *Newtonian liquids*. If v_1, v_2 are the velocities of C, B respectively in Fig. 7.3, and h is their distance apart, the velocity gradient between the liquids is defined as $(v_1 - v_2)/h$. The velocity gradient can thus be expressed in (m/s)/m, or as 's^{-1}'.

Thus if A is the area of the liquid surface considered, the frictional force F on the surface is given by

$$F \propto A \times \text{velocity gradient,}$$

or $\qquad F = \eta A \times \text{velocity gradient,}$.   .   .   (1)

where η is a constant of the liquid known as the *coefficient of viscosity*. This expression for the frictional force in a liquid should be contrasted with the case of solid friction, in which the frictional force is independent of the area of contact and of the relative velocity between the solid surfaces concerned (p. 202).

Definition, Units, and Dimensions of Coefficient of Viscosity

The magnitude of η is given by

$$\eta = \frac{F}{A \times \text{velocity gradient}}.$$

The unit of F is a newton, the unit of A is m², and the unit of velocity

SOLID FRICTION, VISCOSITY 205

gradient is 1 m/s per m. Thus η may be defined as *the frictional force per unit area of a liquid when it is in a region of unit velocity gradient.*

The 'unit velocity gradient' = 1 m s^{-1} change per m. Since the 'm' cancels, the 'unit velocity gradient' = 1 per second. From $\eta = F/(A \times \text{velocity gradient})$, it follows that η may be expressed in units of N s m^{-2}, or 'dekapoise'.

The coefficient of viscosity of water at 10°C is 1.3×10^{-3} N s m^{-2}. Since $F = \eta A \times$ velocity gradient, the frictional force on an area of 10 cm^2 in water at 10°C between two layers of water 0.1 cm apart which move with a relative velocity of 2 cm s^{-1} is found as follows:

Coefficient of viscosity $\eta = 1.3 \times 10^{-3}$ N s m^{-2}, $A = 10 \times 10^{-4}$ m^2, velocity gradient = 2×10^{-2} m s$^{-1} \div 0.1 \times 10^{-2} = 2/0.1$ s^{-1}.

$$\therefore F = 1.3 \times 10^{-3} \times 10 \times 10^{-4} \times 2/0.1 = 2.6 \times 10^{-5} \text{ newton.}$$

Dimensions. The dimensions of a force, F, (= mass × acceleration = mass × velocity change/time) are MLT^{-2}. See p. 14. The dimensions of an area, A, are L^2. The dimensions of velocity gradient

$$= \frac{\text{velocity change}}{\text{distance}} = \frac{L}{T} \div L = \frac{1}{T}.$$

Now
$$\eta = \frac{F}{A \times \text{velocity gradient}},$$

$$\therefore \text{dimensions of } \eta = \frac{\text{MLT}^{-2}}{\text{L}^2 \times 1/\text{T}},$$

$$= \text{ML}^{-1}\text{T}^{-1}.$$

Thus η may be expressed in units 'kg m^{-1} s^{-1}'.

Steady Flow of Liquid Through Pipe. Poiseuille's Formula

The steady flow of liquid through a pipe was first investigated thoroughly by POISEUILLE in 1844, who derived an expression for the volume of liquid issuing per second from the pipe. The proof of the formula is given on p. 208, but we can derive most of the formula by the *method of dimensions* (p. 34).

The volume of liquid issuing per second from the pipe depends on (i) the coefficient of viscosity, η, (ii) the radius, a, of the pipe, (iii), the *pressure gradient*, g, set up along the pipe. The pressure gradient = p/l, where p is the pressure difference between the ends of the pipe and l is its length. Thus x, y, z being indices which require to be found, suppose

$$\text{volume per second} = k\eta^x a^y g^z . \quad . \quad . \quad (1)$$

Now the dimensions of volume per second are L^{3}T^{-1}; the dimensions of η are $ML^{-1}T^{-1}$, see p. 205; the dimension of a is L; and the dimensions of g are

$$\frac{[\text{pressure}]}{[\text{length}]}, \text{ or } \frac{[\text{force}]}{[\text{area}][\text{length}]}, \text{ or } \frac{MLT^{-2}}{L^2 \times L}, \text{ which is } ML^{-2}T^{-2}.$$

Thus from (i), equating dimensions on both sides,

$$L^3 T^{-1} \equiv (ML^{-1}T^{-1})^x L^y (ML^{-2}T^{-2})^z.$$

Equating the respective indices of M, L, T on both sides, we have

$$x + z = 0,$$
$$-x + y - 2z = 3,$$
$$x + 2z = 1.$$

Solving, we obtain $x = -1$, $z = 1$, $y = 4$. Hence, from (1),

$$\text{volume per second} = k\frac{a^4 g}{\eta} = k\frac{pa^4}{l\eta}.$$

We cannot obtain the numerical factor k from the method of dimensions. As shown on p. 209, the factor of $\pi/8$ enters into the formula, which is:

$$\textbf{Volume per second} = \frac{\pi \mathbf{p a}^4}{8\eta \mathbf{l}}. \qquad . \qquad . \qquad (2)$$

EXAMPLE

Explain as fully as you can the phenomenon of viscosity, using the viscosity of a gas as the basis of discussion. Show by the method of dimensions how the volume of liquid flowing in unit time along a uniform tube depends on the radius of the tube, the coefficient of viscosity of the liquid, and the pressure gradient along the tube.

The water supply to a certain house consists of a horizontal water main 20 cm in diameter and 5 km long to which is joined a horizontal pipe 15 mm in diameter and 10 m long leading into the house. When water is being drawn by this house only, what fraction of the total pressure drop along the pipe appears between the ends of the narrow pipe? Assume that the rate of flow of the water is very small. (*O. & C.*)

$$\text{Volume per second} = \frac{\pi p a^4}{8\eta l}, \text{ with usual notation.}$$

$$\text{Thus volume per second} = \frac{\pi p_1 \cdot 0 \cdot 1^4}{8\eta \cdot 5 \times 10^3} = \frac{\pi p_2 \cdot 0 \cdot 0075^4}{8\eta \cdot 10}$$

where p_1, p_2 are the respective pressures in the two pipes, since the volume per second is the same.

$$\therefore \frac{p_1}{p_2} = \frac{0.0075^4}{0.1^4} \times \frac{5 \times 10^3}{10} = \frac{1}{63} \text{ (approx.)}.$$

$$\therefore p_2 = \frac{63}{64} \times \text{total pressure} = 0.984 \times \text{total pressure}.$$

Turbulent Motion

Poiseuille's formula holds as long as the velocity of each layer of the liquid is parallel to the axis of the pipe and the flow pattern has been developed. As the pressure difference between the ends of the pipe is increased, a critical velocity is reached at some stage, and the motion of the liquid changes from an orderly to a *turbulent* one. Poiseuille's formula does not apply to turbulent motion.

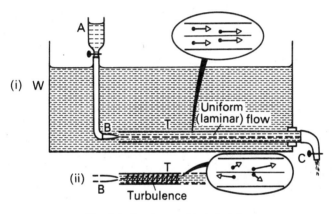

FIG. 7.4 Laminar and turbulent flow

The onset of turbulence was first demonstrated by O. REYNOLDS in 1883, and was shown by placing a horizontal tube T, about 0·5 cm in diameter, at the bottom of a tank W of water, Fig. 7.4 (i). The flow of water along T is controlled by a clip C on rubber tubing connected to T. A drawn-out glass jet B, attached to a reservoir A containing coloured water, is placed at one end of T, and at low velocities of flow a thin coloured stream of water is observed flowing along the middle of T. As the rate of flow of the water along T is increased, a stage is reached when the colouring in T begins to spread out and fill the whole of the tube, Fig. 7.4 (ii). The critical velocity has now been exceeded, and turbulence has begun.

Fig. 7.4 shows diagrammatically in inset: (i) laminar or uniform flow—here particles of liquid at the same distance from the axis always have equal velocities directed parallel to the axis, (ii) turbulence—here particles at the same distance from the axis have different velocities, and these vary in magnitude and direction with time.

Analogy with Ohm's Law

For orderly flow along a pipe, Poiseuille's formula in equation (2) states:

$$\text{Volume per second flowing} = \frac{\pi p a^4}{8\eta l},$$

$$= \frac{p \times \pi a^2}{8\pi\eta \times \dfrac{l}{\pi a^2}}.$$

Now $p \times \pi a^2$ = excess pressure × area of cross-section of liquid = excess force F on liquid, and $l/\pi a^2 = l/A$, where A is the area of cross-section.

$$\therefore \text{volume per second flowing} = \frac{F}{8\pi\eta \times \dfrac{l}{A}}. \qquad (i)$$

The volume of liquid per second is analogous to electric current (I) if we compare the case of electricity flowing along a conductor, and the excess force F is analogous to the potential difference (V) along the conductor. Also, the resistance R of the conductor $= \rho l/A$, where ρ is its resistivity, l is its length, and A is the cross-sectional area. Since, from Ohm's law, $I = V/R$, it follows from (i) that

$$8\pi\eta \text{ is analogous to } \rho, \text{ the resistivity};$$

that is, the coefficient of viscosity η is a measure of the 'resistivity' of a liquid in orderly flow.

Proof of Poiseuille's Formula. Suppose a pipe of radius a has a liquid flowing steadily along it. Consider a cylinder of the liquid of radius r having the same axis as the pipe, where r is less than a. Then the force on this cylinder due to the excess pressure $p = p \times \pi r^2$. We can imagine the cylinder to be made up of cylindrical *shells*; the force on the cylinder due to viscosity is the algebraic sum of the viscous forces on these shells. The force on one shell is given by $\eta A dv/dr$, where dv/dr is the corresponding velocity gradient and A is the surface area of the shell. And although dv/dr changes as we proceed from the narrowest shell outwards, the forces on the neighbouring shells cancel each other out, by the law of action and reaction, leaving a net force of $\eta A dv/dr$, where dv/dr is the velocity gradient at the surface of the cylinder. The viscous force on the cylinder, and the force on it due to the excess pressure p, are together zero since there is no acceleration of the liquid, i.e., we have orderly or laminar flow.

SOLID FRICTION, VISCOSITY

$$\therefore \eta A \frac{dv}{dr} + \pi r^2 p = 0.$$

$$\therefore \eta \cdot 2\pi r l \frac{dv}{dr} + \pi r^2 p = 0, \text{ since } A = 2\pi r l.$$

$$\therefore \frac{dv}{dr} = -\frac{pr}{2\eta l}.$$

$$\therefore v = -\frac{p}{4\eta l} r^2 + c,$$

where c is a constant. Since $v = 0$ when $r = a$, at the surface of the tube, $c = pa^2/4\eta l$.

$$\therefore v = \frac{p}{4\eta l}(a^2 - r^2) \qquad . \qquad . \qquad . \qquad . \qquad (i)$$

Consider a cylindrical shell of the liquid between radii r and $(r + \delta r)$. The liquid in this shell has a velocity v given by the expression in (i), and the volume per second of liquid flowing along this shell = $v \times$ cross-sectional area of shell, since v is the distance travelled in one second, $= v \times 2\pi r \cdot \delta r$.

$$\therefore \text{ total volume of liquid per second along tube } = \int_0^a v \cdot 2\pi r \cdot dr$$

$$= \int_0^a \frac{p}{4\eta l}(a^2 - r^2) \cdot 2\pi r \cdot dr$$

$$= \frac{\pi p a^4}{8\eta l}.$$

EXAMPLE

Define coefficient of viscosity and explain how it may be measured for a viscous liquid such as treacle.

A vertical cylinder of internal diameter 1·02 cm is filled with glycerine. Calculate the velocity attained by a thin-walled brass tube 1·00 cm in diameter, weighing 2·5 g/cm length, which is falling coaxially down the cylinder. Neglect end effects. (Viscosity of glycerine = 0·83 N s m^{-2}; g = 9·8 m s^{-2}.) (C.S.)

Suppose T is the falling hollow brass tube of length l m. Fig. 7.5. Then, neglecting buoyancy, downward force on cylinder = weight = constant viscous force F on liquid between T and cylinder.

Now $$F = -\eta A \frac{dv}{dr},$$

where A is the surface area of a column of glycerine at a distance r from the centre and dv/dr is the velocity gradient there. But $A = 2\pi r l$.

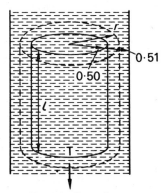

FIG. 7.5 Example

$$\therefore F = -\eta \cdot 2\pi r l \frac{dv}{dr}.$$

$$\therefore \int_{v=u}^{v=0} dv = \frac{F}{\eta \cdot 2\pi l} \int_{0.0050}^{0.0051} -\frac{dr}{r},$$

where u is the velocity attained by the tube and v is the velocity of the glycerine at a distance r from the axis. Since $F = 2 \cdot 5 \times 10^{-3} \times l \times 100g$ newton

$$\therefore u = \frac{2 \cdot 5 \times 0 \cdot 98}{0 \cdot 83 \times 2\pi} \log_e \left(\frac{51}{50}\right)$$

$$= 9 \cdot 3 \times 10^{-3} \text{ m s}^{-1} \text{ (approx.)}.$$

Determination of Viscosity by Poiseuille's Formula

The viscosity of a liquid such as water can be measured by connecting one end of a capillary tube T to a constant pressure apparatus A, which provides a *steady* flow of liquid, Fig. 7.6. By means of a beaker B

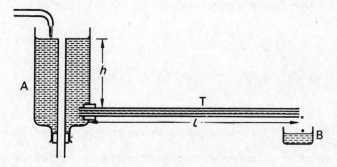

FIG. 7.6 Absolute measurement of viscosity

and a stop-clock, the volume of water per second flowing through the tube can be measured. The pressure difference between the ends of T is $h\rho g$, where h is the pressure head, ρ is the density of the liquid, and g is $9 \cdot 8$ m s^{-2}.

$$\therefore \text{volume per second} = \frac{\pi p a^4}{8 \eta l} = \frac{\pi h \rho g a^4}{8 \eta l},$$

where l is the length of T and a is its radius. The radius of the tube can be measured by means of a mercury thread or by a microscope. The coefficient of viscosity η can then be calculated, since all the other quantities in the above equation are known.

Comparison of Viscosities. Ostwald Viscometer

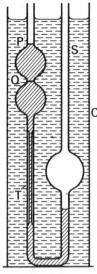

Fig. 7.7
Ostwald viscometer

An Ostwald viscometer, which contains a vertical capillary tube T, is widely used for comparing the viscosities of two liquids, Fig. 7.7. The liquid is introduced at S, drawn by suction above P, and the time t_1 taken for the liquid level to fall between the fixed marks P, Q is observed. The experiment is then repeated with the *same volume* of a second liquid, and the time t_2 for the liquid level to fall from P to Q is noted.

Suppose the liquids have respective densities ρ_1, ρ_2. Then, since the average head h of liquid forcing it through T is the same in each case, the pressure excess between the ends of T = $h\rho_1 g$, $h\rho_2 g$ respectively. If the volume between the marks P, Q is V, then, from Poiseuille's formula, we have

$$\frac{V}{t_1} = \frac{\pi(h\rho_1 g)a^4}{8\eta_1 l} \qquad . \qquad . \qquad \text{(i)},$$

where a is the radius of T, η_1 is the coefficient of viscosity of the liquid, and l is the length of T. Similarly, for the second liquid,

$$\frac{V}{t_2} = \frac{\pi(h\rho_2 g)a^4}{8\eta_2 l} \qquad . \qquad . \qquad . \qquad \text{(ii)}$$

Dividing (ii) by (i),

$$\therefore \frac{t_1}{t_2} = \frac{\eta_1 \rho}{\eta_2 \rho_1}.$$

$$\therefore \frac{\eta_1}{\eta_2} = \frac{t_1}{t_2} \cdot \frac{\rho_1}{\rho_2} \qquad . \qquad . \qquad . \qquad \text{(iii)}$$

Thus knowing t_1, t_2 and the densities ρ_1, ρ_2, the coefficients of viscosity can be compared. Further, if a pure liquid of a known viscosity is used, the viscometer can be used to measure the coefficient of viscosity of a liquid. Since the viscosity varies with temperature, the viscometer should be used in a cylinder C and surrounded by water at a constant temperature, Fig. 7.7. The arrangement can then also be used to investigate the variation of viscosity with temperature. In very accurate work a small correction is required in equation (iii). BARR, an authority on viscosity, estimates that nearly 90% of petroleum oil is tested by an Ostwald viscometer.

Experiment shows that the viscosity coefficient of a liquid diminishes as its temperature rises. Thus for water, η at 15°C is $1 \cdot 1 \times 10^{-3}$ N s m^{-2}, at 30°C it is $0 \cdot 8 \times 10^{-3}$ N s m^{-2} and at 50°C it is $0 \cdot 6 \times 10^{-3}$ N s m^{-2}. Lubricating oils for motor engines which have the same coefficient of viscosity in summer and winter are known as 'viscostatic' oils.

Stokes' Law. Terminal Velocity

When a small object, such as a steel ball-bearing, is dropped into a viscous liquid like glycerine it accelerates at first, but its velocity soon reaches a steady value known as the *terminal velocity*. In this case the viscous force acting upwards, and the upthrust due to the liquid on the object, are together equal to its weight acting downwards, so that the resultant force on the object is zero. An object dropped from an aeroplane at first increases its speed v, but soon reaches its terminal speed. Fig. 7.8 shows that variation of v with time as the terminal velocity v_0 is reached.

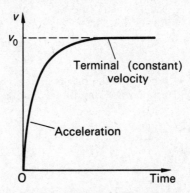

FIG. 7.8 Motion of falling sphere

Suppose a sphere of radius a is dropped into a viscous liquid of coefficient of viscosity η, and its velocity at an instant is v. The frictional force, F, can be partly found by the method of dimensions. Thus suppose $F = ka^x\eta^y v^z$, where k is a constant. The dimensions of F are MLT^{-2}; the dimension of a is L; the dimensions of η are ML^{-1}T^{-1}; and the dimensions of v are LT^{-1}.

$$\therefore \text{MLT}^{-2} \equiv \text{L}^x \times (\text{ML}^{-1}\text{T}^{-1})^y \times (\text{LT}^{-1})^z.$$

Equating indices of M, L, T on both sides,

$$\therefore y = 1,$$
$$x - y + z = 1,$$
$$-y - z = -2.$$

Hence $z = 1$, $x = 1$, $y = 1$. Consequently $F = k\eta a v$. In 1850 STOKES showed mathematically that the constant k was 6π, and he arrived at the formula

$$F = 6\pi a \eta v \qquad . \qquad . \qquad . \qquad . \qquad (1)$$

Comparison of Viscosities of Viscous Liquids

Stokes' formula can be used to compare the coefficients of viscosity of very viscous liquids such as glycerine or treacle. A tall glass vessel G is filled with the liquid, and a small ball-bearing P is dropped gently into the liquid so that it falls along the axis of G, Fig. 7.9. Towards the middle of the liquid P reaches its terminal velocity v_0, which is measured by timing its fall through a distance AB or BC.

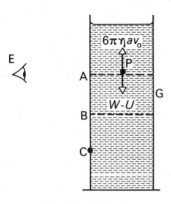

FIG. 7.9 Stokes' law

The upthrust, U, on P due to the liquid $= 4\pi a^3 \sigma g/3$, where a is the radius of P and σ is the density of the liquid. The weight, W, of P is $4\pi a^3 \rho g/3$, where ρ is density of the bearing's material. The net downward force is thus $4\pi a^3 g(\rho - \sigma)/3$. When the opposing frictional force grows to this magnitude, the resultant force on the bearing is zero. Thus for the terminal velocity v_0, we have

$$6\pi \eta a v_0 = \tfrac{4}{3}\pi a^3 g(\rho - \sigma),$$

$$\therefore \eta = \frac{2ga^2(\rho - \sigma)}{9v_0} \qquad . \qquad . \qquad . \qquad (i)$$

When the experiment is repeated with a liquid of coefficient of viscosity η_1 and density σ_1, using the same ball-bearing, then

$$\eta_1 = \frac{2ga^2(\rho - \sigma_1)}{9v_1} \qquad . \qquad . \qquad . \qquad (ii)$$

where v_1 is the new terminal velocity. Dividing (i) by (ii),

$$\therefore \frac{\eta}{\eta_1} = \frac{v_1(\rho - \sigma)}{v_0(\rho - \sigma_1)} \qquad . \qquad . \qquad . \qquad (iii)$$

Thus knowing v_1, v, ρ, σ_1, σ, the coefficients of viscosity can be compared. In very accurate work a correction to (iii) is required for the effect of the walls of the vessel containing the liquid.

EXAMPLE

Define *coefficient of viscosity* and find its dimensions. State *Stokes' law* for the viscous force on a sphere of radius r moving with uniform velocity v through a medium for which the coefficient of viscosity is η. Explain briefly why this law is applicable only if v is less than some critical value v_c. If v_c is of the form $v_c = k\eta^x \rho^y r^z$, where k is a dimensionless constant and ρ is the density of the medium, use the method of dimensions to find the values of x, y and z.

A steel sphere of radius 0·50 cm and mass 4·00 g is released from rest inside a large volume of oil of coefficient of viscosity value $1·20$ N s m^{-2}. Assuming that Stokes' law can be applied even if the velocity is varying, write down the equation of motion of the sphere and derive an equation relating the time after release to the velocity acquired. Hence determine how soon after release the sphere will have acquired a velocity within 1% of the terminal velocity ($\log_e 10 = 2·303$). (*L*.)

(i) From
$$v_c = k\eta^x \rho^y r^z,$$
then, since dimensions are as follows: $v_c = LT^{-1}$, $\eta = ML^{-1}T^{-1}$, $\rho = ML^{-3}$, $r = L$, we have
$$LT^{-1} = (ML^{-1}T^{-1})^x (ML^{-3})^y (L)^z.$$

Equating dimensions of M, L, T on both sides,
$$\therefore x+y = 0; \; -x-3y+z = 1; \; -x = -1.$$
$$\therefore x = 1, y = -1, z = -1.$$

Thus
$$v_c = k\eta/\rho r.$$

(ii) Net force on sphere, $F =$ weight $-$ upthrust $- 6\pi\eta av$.
$$\therefore F = m'g - 6\pi\eta av,$$
where m' is the mass of the sphere less the mass of an equal volume of liquid.

Now
$$F = ma = m\frac{dv}{dt}$$

$$\therefore m\frac{dv}{dt} = m'g - 6\pi\eta av$$

$$\therefore \int_0^t dt = \int_0^v \frac{m\,dv}{m'g - 6\pi\eta av}$$

$$\therefore t = \frac{m}{6\pi\eta a} \log_e\left(\frac{m'g}{m'g - 6\pi\eta av}\right) \qquad . \qquad . \qquad . \qquad (1)$$

At the terminal velocity v_0, $m'g = 6\pi\eta a v_0$.

When v is within 1% of v_0, then $6\pi\eta av = 99\%$ of $m'g = 0·99\,m'g$.

From (1),
$$\therefore t = \frac{m}{6\pi\eta a} \log_e\left(\frac{m'g}{0·01 m'g}\right) = \frac{m}{6\pi\eta a} \log_e 100$$

Substituting $m = 4 \times 10^{-3}$, $\eta = 1\cdot2$, $a = 0\cdot5 \times 10^{-2}$,

$$\therefore t = \frac{4 \times 10^{-3}}{6\pi \times 1\cdot2 \times 0\cdot5 \times 10^{-2}} 2 \log_e 10$$

$$= \frac{4 \times 2 \times 2\cdot303}{6\pi \times 12 \times 0\cdot5} = 0\cdot16 \text{ second.}$$

(Note that the sphere reaches practically its terminal velocity a short time after it is dropped into the liquid.)

Viscosity of Liquid by Rotating Cylinder. The viscosity of a liquid can be measured by means of a cylinder rotating at constant speed ω_0 about its central axis, a method due to Searle. The fixed outer cylinder, A, contains the liquid, and a smaller coaxial cylinder B, pivoted about its central axis, is turned by string round a drum P attached to two equal falling weights, which provide a couple of constant moment G. Fig. 7.10.

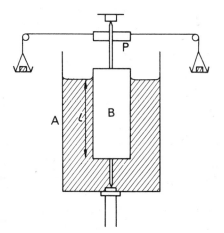

FIG. 7.10 Viscosity of liquid

The angular velocity of the liquid between B and the surface of A varies from ω_0 to zero. Since the velocity v at a distance r from the central axis is $r\omega$, the velocity gradient, dv/dr, $= r d\omega/dr$. Consider now a coaxial cylindrical shell of the liquid between radii r and $r + \delta r$. Since the frictional force F acts over a surface area $2\pi rl$, where l is the depth of the bottom of B below the surface,

$$F = \eta A \frac{dv}{dr} = \eta . 2\pi rl \, r \frac{d\omega}{dr}.$$

Now moment of F about central axis = couple $G = F.r$.

$$\therefore G = \eta . 2\pi r^3 . \frac{d\omega}{dr}$$

$$\therefore \frac{G}{2\pi \eta l} \int_a^b \frac{dr}{r^3} = \int_0^{\omega_0} d\omega,$$

$$\therefore \frac{G}{2\pi \eta l}\left(\frac{1}{2a^2} - \frac{1}{2b^2}\right) = \omega_0,$$

$$\therefore G = \frac{4\pi \eta l a^2 b^2 \omega_0}{b^2 - a^2},$$

or

$$\eta = \frac{G(b^2 - a^2)}{4\pi l a^2 b^2 \omega_0}.$$

Since the couple, G, $= mgd$, where m is the total mass on each scale-pan and d is the diameter of the wheel P, η can be found when ω_0 is determined and the other quantities are measured.

This calculation has assumed stream-line motion at the lower end of B, and omitted the viscous and other forces at the bottom of the inner cylinder. If the total effect on B is equivalent to a couple of moment $c\omega_0$, where c is some constant, then, more accurately,

$$G = \left(\frac{4\pi\eta la^2b^2}{b^2-a^2}+c\right)\omega_0.$$

The effect of c can be eliminated by using two different depths l_1, l_2 of liquid, and arranging the weights on the scale-pans for equilibrium when the angular velocity is ω_0 in each case. Then, by subtraction,

$$G_1-G_2 = \frac{4\pi\eta(l_1-l_2)a^2b^2\omega_0}{b^2-a^2}.$$

Several types of viscometers have been developed on the rotating cylinder principle. In some, the inner cylinder is fixed and the outer cylinder driven by a motor. In 1951 Boyle designed a viscometer which basically uses the rotor of a small motor as a rotating inner cylinder, and the field or stator assembly as the fixed outer cylinder, with the liquid between the two. When the motor is working at a steady low speed, the power developed is a function of the coefficient of viscosity of the liquid, which can thus be read from a calibrated electrical meter in the circuit. The Boyle viscometer is used to investigate the variation of viscosity of liquid at high pressure such as oil in pipe-lines.

Molecular theory of viscosity

Viscous forces are detected in gases as well as in liquids. Thus if a disc is spun round in a gas close to a suspended stationary disc, the latter rotates in the same direction. The gas hence transmits frictional forces. The flow of gas through pipes, particularly in long pipes as in transmission of natural gas from the North Sea area, is affected by the viscosity of the gas.

The viscosity of *gases* is explained by the transfer of momentum which takes place between neighbouring layers of the gas as it flows in a particular direction. Fast-moving molecules in a layer X cross with their own velocity to a layer Y say where molecules are moving with a slower velocity. Fig. 7.11. Molecules in Y likewise move to X. The net

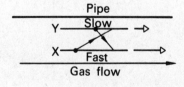

FIG. 7.11   Viscosity of gas–momentum effect

effect is an increase in momentum in Y and a corresponding decrease in X, although on the average the total number of molecules in the two

SOLID FRICTION, VISCOSITY

layers is unchanged. Thus the layer Y speeds up and the layer X slows down, that is, a *force* acts on the layers of the gas while they move. This is the viscous force. We consider the movement of molecules in more detail shortly.

Although there is transfer of momentum as in the gas, the viscosity of a *liquid* is mainly due to the molecular attraction between molecules in neighbouring layers. Energy is needed to drag one layer over the other against the force of attraction. Thus a shear stress is required to make the liquid move in laminar flow.

Formula for Viscosity on Kinetic Theory. On the kinetic theory of gases, the viscosity or frictional force is accounted for by the transfer of momentum across layers of the gas while it is flowing. Fast moving layers lose molecules to slower-moving layers, and vice-versa, so that changes of momentum take place continually across a given layer, and a corresponding force is produced on it.

Viscosity formula. As a simple example, suppose a gas is moving in a given direction Oz. Then $n/6$ is the number of molecules per unit volume moving normally across this direction, along Ox say, where n is the number of molecules per unit volume. If the average velocity of a molecule is v, the number crossing an area A per second $= nAv/6$. On the average, the molecules crossing a given plane come from two planes on either side each a distance λ away, where λ is the mean free path of the molecules. The molecules in one plane have a velocity $v + \lambda dv/dx$, and the molecules in the other plane have a velocity $v - \lambda dv/dx$, where dv/dx represents the velocity gradient in the direction Ox perpendicular to Oz.

$$\therefore \text{ momentum change per second} = \frac{nmAv}{6}\left[\left(v + \lambda\frac{dv}{dx}\right) - \left(v - \lambda\frac{dv}{dx}\right)\right]$$

$$\therefore \text{ frictional force, } F, = \tfrac{1}{3}nm\lambda vA\frac{dv}{dx}.$$

But
$$F = \eta A\frac{dv}{dx}$$

$$\therefore \eta = \tfrac{1}{3}nm\lambda v = \tfrac{1}{3}\rho\lambda v \qquad . \qquad . \qquad . \qquad . \quad \text{(i)}$$

where ρ is the density of the gas.

Mean free path formula. If σ is the effective diameter of a molecule moving with a velocity c in a constant direction, it will make collisions with all molecules whose distance on either side of its centre is σ or less. In one second, the volume of the cylinder containing these molecules is hence $\pi\sigma^2 c$, and thus the number of collisions made is $\pi\sigma^2 cn$, where n is the number of molecules per unit volume.

$\therefore$ average distance between collisions = mean free path λ

$$= \frac{\text{distance moved per second}}{\text{number of collisions}}$$

$$= \frac{c}{\pi\sigma^2 cn} = \frac{1}{\pi\sigma^2 n} \qquad . \qquad . \qquad . \qquad . \quad \text{(ii)}$$

This is an approximate formula for λ; more accurately, Maxwell showed that $\lambda = 1/\sqrt{2}\pi\sigma^2 n$. Thus $\lambda \propto 1/n$. Now the number of molecules per unit volume, n, is proportional to the pressure of the gas. Hence $\lambda \propto 1/p$. From the expression for η in (i), it can now be seen that η *is independent of the pressure*, since ρ, the density, is proportional to pressure. This surprising result was verified by Maxwell by experiment, and it helps to confirm the general truth of the kinetic theory of gases.

Viscosity of a Gas. On p. 206 it was shown that the volume per second, v, of liquid flowing along a tube under stream-line motion was given by

$$v = \frac{\pi p a^4}{8\eta l}.$$

In deriving the formula it was assumed that the volume per second crossing each section of the tube was constant, which is true for an incompressible substance and hence fairly true for a liquid. When a *gas* flows along a tube, however, the volume increases as the pressure decreases, and hence Poiseuille's formula above must be modified to take this into account.

For a short length δl of the tube, the velocity can be considered constant. The small chance of pressure across this length is δp, and the pressure gradient is thus $-dp/dl$, the minus indicating that the pressure diminishes as l increases. Poiseuille's formula now becomes

$$v = \frac{-\pi a^4}{8\eta} \cdot \frac{dp}{dl} \quad . \quad . \quad . \quad . \quad (i)$$

But from Boyle's law, $pv = P_1 V_1 = P_2 V_2$, where P_1, P_2 are the respective pressures at the inlet and outlet of the tube, and V_1, V_2 are the corresponding volumes per second. Thus $v = P_1 V_1 / p$. Substituting in (i),

$$\therefore \frac{P_1 V_1}{p} = -\frac{\pi a^4}{8\eta} \cdot \frac{dp}{dl}$$

$$\therefore P_1 V_1 \int_0^l dl = -\frac{\pi a^4}{8\eta} \int_{P_1}^{P_2} p \, dp.$$

$$\therefore P_1 V_1 = \frac{\pi a^4}{16 \eta l}(P_1^2 - P_2^2) = P_2 V_2 \quad . \quad . \quad (ii)$$

A simple method of measuring the *viscosity of air* is illustrated in Fig. 7.12. A tube HL of a few millimetres diameter is joined to a fine capillary tube T, and a mercury pellet M is introduced at the top, as shown. The time taken for M to fall a measured height HL is noted. During this time a volume of air equal to that between H, L is driven through T, and hence the volume per second V_1 is known if the diameter of HL is measured. The pressure P_2 at the open end of T is atmospheric pressure, A; the pressure P_1 at the other end is $(A+p)$, where p is the pressure due to the pellet of mercury Since $p = mg/b$, where m is the mass of the pellet and b the cross-sectional area of HL, p can be evaluated. Thus knowing the length l and radius a of the capillary tube T, the viscosity η can be found by substituting in equation (ii). A correction is necessary as the mercury sticks to the side of the tube. In Rankine's method, a completely closed tube system is used. Details must be obtained from more advanced works.

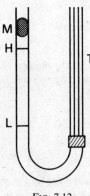

FIG. 7.12
Viscosity of air

SOLID FRICTION, VISCOSITY

EXERCISE 7

What are the missing words in the statements 1–6?

1. The coefficient of dynamic (kinetic) friction is the ratio . . .

2. The coefficient of friction between two given surfaces is . . . of the area in contact.

3. In orderly or laminar flow of liquids in a pipe, the volume per second flowing past any section is given by the formula . . .

4. The dimensions of coefficient of viscosity are . . .

5. When a small sphere of radius a falls through a liquid with a constant velocity v, the frictional force is given by the formula . . .

6. In comparing the viscosities of water and alcohol by an Ostwald viscometer, the same liquid . . . must used.

Which of the following answers, A, B, C, D or E, do you consider is the correct one in the statements 7–10?

7. In orderly or laminar flow of a liquid through a pipe, A tensile forces act on the layers and the volume per second V is proportional to the pressure at one end, B shear forces act on the layers and V is proportional to the pressure at one end, C shear forces act on the layers and V is proportional to the pressure difference between the ends, D bulk forces act throughout the liquid, E V is directly proportional to a^4 and to the coefficient of viscosity.

8. When a small steel sphere is dropped gently down the axis of a wide jar of glycerine, the sphere A travels with constant velocity throughout its motion, B accelerates at first and then reaches a constant velocity, C decelerates at first and then reaches a constant velocity, D accelerates throughout its motion, E slowly comes to rest.

9. When a gas flows steadily along a pipe, the viscous forces in it are due to A transfer of energy from one layer to another, B the uniform speed of the molecules, C the varying density along the pipe, D the transfer of momentum from one layer to another, E the varying pressure at a given section of the pipe.

10. A pipe P has twice the diameter of a pipe Q, and P has a liquid X flowing along it which has twice the viscosity of a liquid Y flowing through Q. If the flow is orderly or laminar in each, and the volume per second in P and Q is the same, the pressure difference at the ends of P compared to that of Q is A 1:8, B 1:4, C 8:1, D 4:1, E 1:1.

Solid Friction

11. State the laws of solid friction. Describe an experiment to determine the coefficient of dynamic (or sliding) friction between two surfaces.

A horizontal circular turntable rotates about its centre at the uniform rate of 120 revolutions per minute. Find the greatest distance from the centre at which a small body will remain stationary relative to the turntable, if the coefficient of static friction between the turntable and the body is 0·80. (*L*.)

220   MECHANICS AND PROPERTIES OF MATTER

12. State (a) the laws of solid friction, (b) the triangle law for forces in equilibrium. Describe an experiment to determine the coefficient of sliding (dynamic) friction between two wooden surfaces.

A block of wood of mass 150 g rests on an inclined plane. If the coefficient of static friction between the surfaces in contact is 0·30, find (a) the greatest angle to which the plane may be tilted without the block slipping, (b) the force parallel to the plane necessary to prevent slipping when the angle of the plane with the horizontal is 30°, showing that this direction of the force is the one for which the force required to prevent slipping is a minimum. ($g = 10$ m s^{-2}.) (L.)

13. Distinguish between *static* and *sliding* (kinetic) friction and define the *coefficient of sliding friction*.

How would you investigate the laws of sliding friction between wood and iron?

An iron block, of mass 10 kg, rests on a wooden plane inclined at 30° to the horizontal. It is found that the least force parallel to the plane which causes the block to slide *up* the plane is 100 N. Calculate the coefficient of sliding friction between wood and iron. (N.)

14. Give an account of the factors which determine the force of friction (i) between solids, (ii) in liquids.

A block weighing 12 kg is drawn along a horizontal surface by a steadily applied force of 4 kg weight acting in the direction of motion. Find the kinetic energy acquired by the block at the end of 10 seconds and compare it with the total work done on the block in the same time. (Coefficient of friction = 0·28.) (L.)

15. State the laws of solid friction.

Describe experiments to verify these laws, and to determine the coefficient of static friction, for two wooden surfaces.

A small coin is placed on a gramophone turntable at a distance of 7·0 cm from the axis of rotation. When the rate of rotation is gradually increased from zero the coin begins to slide outwards when the rate reaches 60 revolutions per minute. Calculate the rate of rotation for which sliding would commence if (a) the coin were placed 12·0 cm from the axis, (b) the coin were placed in the original position with another similar coin stuck on top of it. (L

16. Define *coefficient of sliding friction, coefficient of viscosity*. Contrast the laws of solid friction with those which govern the flow of liquids through tubes.

Sketch the apparatus you would employ to determine the coefficient of sliding friction between a wood block and a board and show how you would deduce the coefficient from a suitable graph. (L.)

Viscosity

17. Define *coefficient of viscosity* of a fluid.

When the flow is orderly the volume V of liquid which flows in time t through a tube of radius r and length l when a pressure difference p is maintained between its ends is given by the equation $\dfrac{V}{t} = \dfrac{\pi p r^4}{8 l \eta}$ where η is the coefficient of viscosity of the liquid. Describe an experiment based on this equation *either* (a) to determine the value of η for a liquid, *or* (b) to compare the values of η for two liquids, pointing out the precautions which must be taken in the experiment chosen to obtain an accurate result.

SOLID FRICTION, VISCOSITY 221

Water flows steadily through a horizontal tube which consists of two parts joined end to end; one part is 21 cm long and has a diameter of 0·225 cm and the other is 7·0 cm long and has a diameter of 0·075 cm. If the pressure difference between the ends of the tube is 14 cm of water find the pressure difference between the ends of each part. (L.)

18. The dimensions of *energy*, and also those of *moment of a force* are found to be 1 in *mass*, 2 in *length* and −2 in *time*. Explain and justify this statement.

(a) A sphere of radius a moving through a fluid of density ρ with *high* velocity V experiences a retarding force F given by $F = k \cdot a^x \cdot \rho^y \cdot V^z$, where k is a non-dimensional coefficient. Use the method of dimensions to find the values of x, y and z.

(b) A sphere of radius 2 cm and mass 100 g, falling vertically through air of density 1·2 kg m^{-3}, at a place where the acceleration due to gravity is 9·81 m s^{-2}, attains a steady velocity of 30 m s^{-1}. Explain why a constant velocity is reached and use the data to find the value of k in this case. (O. & C.)

19. Mass, length and time are *fundamental* units, whereas acceleration, force and energy are *derived* units. Explain the distinction between these two types of unit. Define each of the three derived units and apply your definition in each case to deduce its dimensions.

An incompressible fluid of viscosity η flows along a straight tube of length l and uniform circular cross-section of radius r. Provided the pressure difference p between the ends of the tube is not too great the velocity u of fluid flow along the axis of the tube is found to be directly proportional to p. Apply the method of dimensions to deduce this result assuming u depends only on r, l, η and p.

How may the viscosity of an ideal gas be accounted for by elementary molecular theory? (O. & C.)

20. Define *coefficient of viscosity*. Describe an experiment to compare the coefficients of viscosity of water and benzene at room temperature.

A small metal sphere is released from rest in a tall wide vessel of liquid. Discuss the forces acting on the sphere (a) at the moment of release, (b) soon after release, (c) after the terminal velocity has been attained.

Castor oil at 20°C has a coefficient of viscosity 2·42 N s m^{-2} and a density 940 kg m^{-3}. Calculate the terminal velocity of a steel ball of radius 2·0 mm falling under gravity in the oil, taking the density of steel as 7800 kg m^{-3}. (L.)

21. Define *coefficient of viscosity*. Distinguish between orderly and turbulent flow of a liquid through a tube. Describe a method to determine for a given tube and liquid the pressure head at which the transition from orderly to turbulent flow occurs.

A horizontal capillary tube, 50 cm long and 0·20 mm internal radius, is inserted into the lower end of a tall cylindrical vessel of cross-sectional area 10 cm^2. The vessel is filled with water which is allowed to flow out through the tube. Calculate the time taken for the level of the water in the vessel to fall from a height of 100 cm to 50 cm above the axis of the tube. Assume that the volume of water passing per second through a horizontal tube is $\pi a^4 (p_1 - p_2)/8l\eta$, where a = tube radius, l = tube length, η = coefficient of viscosity of water, and $(p_1 - p_2)$ = difference in the pressures at the ends of the tube. Take the viscosity of water as $1·0 \times 10^{-3}$ N s m^{-2} and $\log_e 10 = 2·30$. (N.)

22. Define *coefficient of viscosity*. What are its dimensions?

By the method of dimensions, deduce how the rate of flow of a viscous liquid through a narrow tube depends upon the viscosity, the radius of the tube, and the pressure difference per unit length. Explain how you would use your results to compare the coefficients of viscosity of alcohol and water. (*C*.)

23. Define *coefficient of viscosity*. For orderly flow of a given liquid through a capillary tube of length l, radius r, the volume of liquid issuing per second is proportional to pr^4/l where p is the pressure difference between the ends of the tube. How would you verify this relation experimentally for water at room temperature? How would you detect the onset of turbulence? (*N*.)

24. The viscous force acting on a small sphere of radius a moving slowly through a liquid of viscosity η with velocity v is given by the expression $6\pi\eta av$. Sketch the general shape of the velocity-time graph for a particle falling from rest through a viscous fluid, and explain the form of the graph. List the observations you would make to determine the coefficient of viscosity of the fluid from the motion of the particle.

Some particles of sand are sprinkled on to the surface of the water in a beaker filled to a depth of 10 cm. Estimate the least time for which grains of diameter 0·10 mm remain in suspension in the water, stating any assumptions made.

[Viscosity of water = $1·1 \times 10^{-3}$ N s m^{-2}; density of sand = 2200 kg m^{-3}.] (*C*.)

25. Define *coefficient of friction* and *coefficient of viscosity*.

Describe how you would (*a*) measure the coefficient of sliding friction between iron and wood, and (*b*) compare the viscosities of water and paraffin oil. (*L*.)

26. Define *coefficient of viscosity* and deduce its dimensions.

The annular space between an outer fixed cylinder of radius a_1 and an inner coaxial rotatable cylinder of radius a_2 is filled with a liquid of coefficient of viscosity η. If $a_1 - a_2$ is small compared with a_1, find the couple required to cause the inner cylinder to rotate with angular velocity ω when immersed to a depth l. Explain how the effects of the ends of the cylinders can be eliminated in practice. (*C*.)

REVISION PAPER A

(Assume $g = 10$ m s^{-2} unless otherwise specified)

1. State Newton's second law of motion. Show that a force must act on a particle moving in a circle with constant speed, and derive an expression for this force.

What is the nature of the force maintaining the motion of (a) an artificial satellite, (b) an electron in a hydrogen atom? Observations are made of the period of revolution and height of an artificial satellite revolving about the earth in a circular orbit. Show how an estimate of the mass of the earth may be made by using the observations and other necessary data. (*N.*)

2. Define *surface tension* and *angle of contact*.

The densities of water and paraffin oil, which are 1000 and 850 kg m^{-3} respectively, are being compared in a Hare's apparatus constructed with glass tubes of internal diameter 0·20 cm. If the surface tension and angle of contact with glass for water are $7·2 \times 10^{-2}$ N m^{-1} and $0°$ respectively, and the corresponding values for paraffin oil are $2·7 \times 10^{-2}$ N m^{-1} and $26°$, calculate the height at which paraffin oil will stand in its tube when the water stands 37·0 cm above its level in the containing vessel.

What is the percentage error in the value of the density of the oil obtained from this experiment if the surface tension effects are ignored? (*L.*)

3. Explain what is meant by (a) *constant of gravitation*, (b) *acceleration due to gravity*. Establish the relation between them. Give *two* reasons why the acceleration of a freely falling body at sea level is not constant over the surface of the earth.

At a certain location the readings of a mercury barometer and an aneroid barometer are observed to be the same. When the two instruments are moved to another location the aneroid and mercury barometers read 752·1 mm and 750·0 mm respectively, the readings at the two locations being made at the same temperature. What percentage change in the acceleration due to gravity will account for this? If due to a change in atmospheric pressure the mercury barometer subsequently reads 740·0 mm, what will be the new reading of the aneroid barometer? (*N.*)

4. Describe a method for measuring the value of Young's modulus for a metal wire. What precautions must be observed to obtain an accurate result and what factors limit the accuracy of your method?

A uniform rigid disc is suspended by means of four uniform parallel vertical wires which are clamped so that the plane of the disc is horizontal and each wire is under tension. These wires are each 200 cm long and of cross-sectional area 0·5 mm^2. Three of the wires are made of steel and are attached to points equispaced on the periphery of the disc. The fourth wire is made of brass and is attached to the centre of the disc. When an additional mass of 40 kg is hung from the centre of the disc the wires are each extended by the same amount and the disc remains with its plane horizontal. Calculate the extension produced and the increase in tension in each wire when the 40 kg mass is added. [Young's modulus for steel $= 2·1 \times 10^{11}$ N m^{-2} and for brass $= 9·8 \times 10^{10}$ N m^{-2}.] (*O. & C.*)

5. Explain what is meant by the *velocity gradient* at a point in a fluid.

A small sphere when allowed to fall through a viscous medium eventually acquires a terminal velocity. Explain this qualitatively in terms of the forces concerned and discuss the energy transformations occurring when the terminal velocity has been reached.

A small steel ball-bearing falling through glycerine has a terminal velocity of 2.00 cm s^{-1}. Find the terminal velocity of an air bubble of the same size rising through glycerine at the same temperature. Neglect the weight of air in the bubble and assume that the viscous force on each sphere is proportional to its terminal velocity. (Density of steel = 7700 kg m^{-3}. Density of glycerine = 1260 kg m^{-3}.) (*N*.)

6. Define simple harmonic motion, and show that the total energy (kinetic and potential) of a particle executing simple harmonic motion is proportional (*a*) to the square of the amplitude, (*b*) to the square of the frequency.

The total energy of a particle executing a simple harmonic motion of period 2π seconds is 1.024×10^{-3} J. $\pi/4$ seconds after the particle passes the mid-point of the swing its displacement is $8\sqrt{2}$ cm. Calculate the amplitude of the motion and the mass of the particle. (*O. & C.*)

7. Explain the terms *moment of inertia* and *radius of gyration* about an axis.

State a formula for the period of small oscillation of a rigid pendulum about a given axis in terms of its radius of gyration about a parallel axis through the centre of gravity and the distance of the centre of gravity below the axis of suspension. Describe in detail an experiment to verify this formula using a uniform thin rod capable of being suspended at various positions along its length.

A uniform thin disc and a uniform thin hoop of equal radius are each allowed to perform small oscillations about a horizontal axis. In each case the axis is perpendicular to the plane of the disc or hoop and passes through a point on the circumference. Compare the two periods of oscillation. (*L*.)

REVISION PAPER B

1. Explain what is meant by the *momentum* of a body? Describe an experiment to demonstrate the conservation of linear momentum.

A nucleus, originally at rest, undergoes radioactive decay be emitting an electron of momentum 4.5×10^{-21} kg m s^{-1} and a neutral particle of negligible mass but finite momentum. The nucleus recoils with momentum 5.2×10^{-21} kg m s^{-1} at an angle of $150°$ to the direction of emission of the electron. Find the magnitude and direction of the momentum of the neutral particle. (*C*.)

2. Distinguish between *vector* and *scalar* quantities. Classify *mass, force, energy*.

A particle, mass 5 g, executes simple harmonic motion of amplitude 2 cm and frequency 2 Hz. Draw a graph showing how the force acting on the particle varies with its displacement from the equilibrium position. What are the maximum and minimum values of the kinetic energy of the particle? (*N*.)

3. Define the term *coefficient of viscosity* and point out under what conditions the definition given is applicable.

Describe briefly one experiment in each case to compare the coefficients of viscosity of (*a*) water and benzene, (*b*) castor oil and glycerine, at room temperature. Point out any necessary precautions which should be taken. (*L*.)

4. Explain the terms *stress*, *strain*, *modulus of elasticity* and *elastic limit*. Derive an expression in terms of the tensile force and extension for the energy stored in a stretched rubber cord which obeys Hooke's law.

The rubber cord of a catapult has a cross-sectional area $1\cdot0$ mm^2 and a total unstretched length $10\cdot0$ cm. It is stretched to $12\cdot0$ cm and then released to project a missile of mass $5\cdot0$ g. From energy considerations, or otherwise, calculate the velocity of projection, taking Young's modulus for the rubber as $5\cdot0 \times 10^8$ N m^{-2}. State the assumptions made in your calculation. (*L.*)

5. Explain the meaning of the term *moment of inertia*. Derive an expression for the kinetic energy of a rigid body rotating with angular velocity ω about a fixed axis.

Why in some cases is an engine or motor provided with a flywheel? A flywheel which can turn about a horizontal axis where there is a constant frictional couple is set in motion by means of a mass of $0\cdot5$ kg hung from a light string wrapped around an axle of radius $1\cdot25$ cm. The weight falls from rest and the string unwinds without slipping so that the flywheel makes 10 revolutions before the string becomes detached from the axle. The flywheel then makes a further 50 revolutions before it finally makes to rest. If the velocity of the weight at the instant of detachment is $0\cdot2$ m s^{-1}, find the moment of inertia of the flywheel about its axis of rotation. (*N.*)

6. (*a*) Explain what is meant by the term 'plane laminar motion' applied to a viscous fluid. What factors determine the tangential stress experienced by a plane surface over which a viscous fluid flows in a direction parallel to the surface?

Use the method of dimensions to determine the form of the relation between the volume of liquid flowing steadily in unit time through a cylindrical tube, the pressure gradient between the ends of the tube, the viscosity of the liquid, and the cross-sectional area of the tube. [You may assume that no other factors are important in determining the flow conditions.]

(*b*) A non-viscous liquid of density ρ kg m^{-3} moves with steady streamline flow at M kg s^{-1} through a horizontal tube of uniform cross-sectional area $2A$. What is the speed of the liquid? Why can we assume it to be the same all over the cross-section?

Between X and Y the tube narrows uniformly and gradually to a cross-sectional area A at Y. What is the speed of the liquid at Y, and what is the difference between the static pressures at X and Y? (*O.*)

7. Define *surface tension* and state the unit in which it is usually expressed.

Making reference to your definition, derive an expression for the pressure difference between the inside and outside of a spherical drop of radius r.

What is the effect on the surface tension of water of adding a little detergent? How would you attempt to investigate the effect on the surface tension of various concentrations of detergent? (*N.*)

REVISION PAPER C

1. Distinguish between *mass* and *weight*. A body of mass 5 kg is transferred to the surface of the moon. Calculate (*a*) its weight, (*b*) the force required to give it an acceleration of $1\cdot5$ m s^{-2}, (*c*) the force required to keep it in uniform motion on a horizontal plane with a coefficient of friction $0\cdot5$. Assume that the gravitational acceleration on the moon is $1\cdot8$ m s^{-2}.

A mass of 5 kg and relative density 2·5 is suspended, completely immersed in water, from a spring balance attached to the roof of a lift. What will be the reading of the balance (a) if the lift is moving up with an acceleration of 2·4 m s^{-2}, (b) if the lift is falling freely under gravity? (*O. & C.*)

2. Define *simple harmonic motion* and show how it is related to uniform motion in a circle. Hence, or otherwise, derive an expression for the periodic time of small oscillations of a simple pendulum.

A simple pendulum of effective length 50 cm consists of a small bob of mass 25 g supported by a light thread. Calculate (a) the periodic time oscillation, (b) the tension in the thread as the bob passes through its lowest point, if the amplitude of the oscillation is 4·0 cm. (*L.*)

3. State *Hooke's law* and explain the terms *elastic limit, Young's modulus*.

Draw a labelled diagram of an apparatus which could be used to determine accurately the extension of a long wire suspended vertically and loaded at the free end.

When the load on the wire, which is of length 200 cm and diameter 0·60 mm, is gradually increased from 1·75 kg to 4·25 kg the scale reading on the apparatus changes from 1·78 mm to 2·61 mm. Calculate (a) Young's modulus for the material of the wire, (b) the additional potential energy stored in the wire. (*N.*)

4. Assuming Newton's law of gravitation, derive an expression relating the acceleration of gravity g_h, at a height h with its value g_0, at the earth's surface.

An earth satellite is projected from the South pole into an orbit at a height of 300 km by means of a two stage rocket. The first stage accelerates the satellite vertically upwards until a height of 10 km is reached when the vertical velocity is V_1. This stage is then switched off and the satellite moves on upwards and turns on its side when near to its maximum height. The second stage is then fired to accelerate the satellite into orbit with a final velocity V_2. If the final orbit is circular in a plane through the earth's centre, calculate the values of V_1 and V_2, neglecting air resistance. [Radius of the earth = 6360 km.] (*O. & C.*)

5. State the laws of solid friction.

Describe experiments to verify these laws and explain how to determine the coefficient of dynamic friction for two surfaces.

A children's slide of constant slope is 4·5 m in length. The upper end is 2·5 m vertically above the ground while the lower end is at ground level. If a child starts sliding from rest at the upper end of the slide, find its velocity at the lower end assuming the coefficient of dynamic friction between the child and the slide to be constant at 0·25. What percentage is this final velocity of that which would have been attained if the friction were negligible? (*L.*)

6. Define *surface tension* and state the effect on the surface tension of water of raising its temperature.

Describe an experiment to measure the surface tension of water over the range of temperatures from 20°C to 70°C. Why is the usual capillary rise method unsuitable for this purpose?

Two unequal soap bubbles are formed one on each end of a tube closed in the middle by a tap. State and explain what happens when the tap is opened to put the two bubbles into connection. Give a diagram showing the bubbles when equilibrium has been reached. (*L.*)

7. A flat plate of irregular shape is pierced by a number of small holes, distributed at random, through which a knitting needle can pass easily. Describe how, using the plate and the needle, you would find the acceleration due to gravity and give the theory of the method.

A thin uniform rod swings as a pendulum about a horizontal axis at one end, the periodic time being 1·65 second. If the mass of the rod is 125 g, what is (a) its length, (b) its moment of inertia about the horizontal axis? (N.)

REVISION PAPER D

1. Explain what is meant by the *dimensions* of a physical quantity in mass, length and time. Find the dimensions of *weight, velocity gradient, coefficient of viscosity*.

A small sphere of radius r, falling under gravity through a fluid of coefficient of viscosity η, ultimately attains a steady or terminal velocity v. Apply the method of dimensions to determine how v depends upon r, η, and w, where w is the effective weight of the sphere in the fluid, i.e., the difference between the true weight and the upward thrust due to the displaced fluid. (*O. & C.*)

2. Define G, the universal constant of gravitation, and g, the local acceleration due to gravity. Describe how *one* of these quantities can be measured and show how these two quantities may be used to estimate the mean density of the earth.

The diameter of the planet Mercury is about 0·4 times that of the earth and its mean density is approximately equal to that of the earth. The mean radii of the orbits about the sun of Mercury and the earth are in ratio of approximately 1:2. Show qualitatively that the conditions for retaining an atmosphere are less favourable on Mercury than on the earth. (*O.*)

3. Calculate the percentage loss of kinetic energy of a neutron of mass m when it makes a perfectly elastic head-on collision with a stationary beryllium nucleus of mass $9m$. State the physical principles involved in your calculation. (*N.*)

4. You are provided with a length of elastic cord, a set of known masses, a stop clock and the usual facilities of a physical laboratory. Describe how you would use them to find a value for the acceleration due to gravity and give the theory of the method.

The ends of a light elastic cord of natural length 200 cm and diameter 2 mm are attached to two points 200 cm apart in a horizontal plane. Find to the nearest gram the mass which, when suspended in equilibrium from the mid-point of the cord, produces a depression of 10 cm at this point and calculate the energy which would then be stored in the strained cord. Young's modulus for the material of the cord is 5.0×10^7 N m^{-2}. (*N.*)

5. Derive an expression for the moment of inertia of a uniform circular disc of mass M, radius r, about a central axis perpendicular to its plane. How would you determine this moment of inertia experimentally?

A circular disc of mass 800 g, radius 10 cm, is suspended by a wire through its centre perpendicular to its plane and makes 50 torsional oscillations in 59·8 seconds. When an annulus is placed symmetrically on the disc, the system makes 50 oscillations in 66·4 seconds. Calculate the moment of inertia of the annulus about the axis of rotation. (*N.*)

6. (a) Explain the distinction between fundamental and derived units. If mass length and force were used as fundamental quantities what dimensions would the unit of time have?

(b) Distinguish between the sensitivity and the accuracy of a beam balance. Upon what factors does each of these quantities depend in the case in which the knife edges are coplanar?

(c) In an experiment with a reversible pendulum the following expression occurs:
$$C = (T_1^2 - T_2^2)/(l_1 - l_2).$$
If the errors in l_1 and l_2 are both negligible but T_1 and T_2 are both subject to uncertainties $\pm \Delta T$ what is the maximum permissible value of ΔT if C is to be accurate to 1%? (O. & C.)

7. Describe an experiment to determine the surface tension of a soap solution by a method involving the measurement of the excess pressure inside a soap bubble.

A glass rod of mass 1·21 g is suspended by two parallel vertical threads of equal length from a similar rod which is clamped in a horizontal position. The threads are 4·80 cm apart. When a soap film is formed between the threads and the rods the lower rod is raised up and the threads form into curves. When equilibrium is reached the rods are a distance 10·8 cm apart measured vertically, while the vertical tangents to the threads are a distance 2·04 cm apart measured horizontally. Show that each thread forms a circular arc whose radius depends on the tension in the thread and the surface tension of the soap solution. Hence, by considering the equilibrium of the lower half of the system, or otherwise, deduce a value for the surface tension of the soap solution. The weight of the liquid film may be neglected. (L.)

ANSWERS TO EXERCISES

EXERCISES 1 (p. 35)

1. LT^{-1}.
2. MLT^{-2}.
3. (i) scalar, (ii) vector, (iii) scalar, (iv) vector.
4. mass × velocity.
5. momentum, energy.
6. total linear momentum, external.
7. 1 joule.
8. 10 N (approx.).
9. vector.
10. rate.
11. C.
12. B.
13. D.
14. A.
15. (i) 5 s, (ii) 62·5 m, (iii) 18 m s^{-1}.
16. (i) 4 s, (ii) 20 m, (iii) 10, 10 J.
17. 19·8 km h^{-1}, N. 30·5° W.
18. (i) 100, (ii) 500 J.
19. (i) 5·2 m s^{-1}, loss = 58 J, (ii) 1·2 s^{-1}, loss = 314 J.
20. 26·5 km h^{-1}, S. 41° W., 7·6 km.
21. (i) $1\frac{3}{7}$ m s^{-1}, (ii) $1\frac{3}{7}$ m s^{-2}, (iii) 86 J.
22. 10·5° from vertical.
23. $2E/103$.
24. (a) 10/3 N, (b) 5/9 W, (c) 5/18 W.
25. 167 kgf; 83,330 J.
26. 21,675 m, 3,125 m, 25 s; 465 m s^{-1}.
27. 1/3.
28. (a) ML^2T^{-2}, (b) a: ML^5T^{-2}, b: L^3.
29. 14·4 minutes, 8 km, 37° S. of E.
30. $v/2$ at 60° to initial velocity of first sphere.
31. 2 m s^{-1}.
32. 800 kgf, 22 kW.
33. 70%, 0·023.
34. $22·5 \times 10^4$ N m^{-2}.

EXERCISES 2 (p. 80)

1. centripetal.
2. middle.
3. $g = GM/r^2$.
4. end.
5. $1/\text{distance}^2$.
6. 24 h.
7. B.
8. D.
9. C.
10. D.
11. (i) 2 rad s^{-1}, (ii) 9·6 kgf.
12. (i) 11·8 kgf, (ii) 32°.
13. 42°, 1555 kgf.
14. 22·4, 6·4 kgf.
15. (a) mgl, (b) $\sqrt{2gl}$, (c) $2g$ up, (d) $3\ mg$.
16. Break when stone vertically below point of suspension; 7·7 rad s^{-1}; 122 cm from point below point of suspension.
17. (i) 1/20 s, (ii) 0, $3200\pi^2$ cm s^{-2}, (iii) 80π cm s^{-1}, 0.
18. 101 cm, (i) 0, $2\pi^2$ cm s^{-2}, (ii) 2π cm s^{-1}, 0, (iii) $\sqrt{3}\pi$ cm s^{-1}, π^2 cm s^{-2}.
19. 0·20 s.
20. 1·09 s.
21. 10 m s^{-2}, 4·5 m.
22. (a) 2·5 m s^{-1}, (b) 790 m s^{-2}.
23. 6·3 cm.
26. 1·6 Hz.

27. (a) $4\pi^2$, (b) $2\sqrt{3\pi^2}$, (c) $4\pi^3$; $16\pi^4 mr$.
30. $\frac{1}{2}m\omega^2(a^2 - x^2)$.
32. 4 cm.
33. 3×10^{-7} N.
34. 6.0×10^{24} kg.
35. 9.9 m s^{-2}.
38. 1.4×10^{-5} rad s^{-1}.
39. 9.77 m s^{-2}.
40. 5500 kg m^{-3}.
42. $\rho_1 : \rho_2 = 1 : 2$.
43. 24 hours.
45. $93.3 : 1$.

EXERCISES 3 (p. 107)

1. $\frac{1}{2}I\omega^2$
2. $I\omega$.
3. couple.
4. $2\pi\sqrt{I/mgh}$.
5. D.
6. B.
7. A.
8. E.
9. (i) 2×10^{-4}, (ii) 8×10^{-4} kg m^2.
10. 1.2×10^6 J, 3.8×10^5 kg m^2 s^{-1}.
11. 126 s, 63 rev.
12. 15 J.
13. 3.6 m s^{-2}, 6.0 m s^{-1}.
14. (i) 1.9, (ii) 2.2 rad s^{-1}.
15. 4.0 s.
16. (a) 20 rad s^{-2}, (b) 0.32 N m.
17. 7.3×10^{-4} kg m^2.
18. 4.02×10^{-3} kg m^2.
19. (a) 3.4×10^{-2} kg m^2, (b) 6.7×10^{-3} N m, (c) 137.5 rev, (d) 4.2 J.
20. 3.5 m s^{-1}.
21. 6.2×10^{-4} kg m^2.
23. $1 : 12.5$.
24. 2.8×10^{16} kg.

EXERCISES 4 (p. 141)

1. newton metre.
2. rises.
3. meet.
4. $F \cos \theta$.
5. centre of gravity.
6. weight of object.
7. velocity.
8. B.
9. D.
10. C.
11. 22.6 cm.
12. 690 N, 1390 N.
13. 500 N m; 167 N.
14. (i) 0.65, (ii) 0.60 m.
15. 41.7 N, 41.7 N.
16. $30°$.
18. 1710 N.
19. 0.04 cm; $0.1°$.
20. $\pm \cos^{-1}[(M + m)r \sin \phi / Ml]$.
22. 180 N, 159 N.
23. stable if $r > t/2$.
25. (i) $3 : 2$, (ii) $40 : 9$.
26. 1.62 N.
27. 5.1 cm.
28. $2 : 3$.
29. $W\rho/\sigma$, (a) 0.95, (b) 1.19.
30. (a) 75.0 cm, (b) 1003.3 cm.
33. 1.25.
34. 5.7 cm^3.

ANSWERS   231

EXERCISES 5 (p. 175)

1. $N\,m^{-1}$.
2. MT^{-2}.
3. minimum.
4. $4\gamma/r$.
5. $2\gamma/r$.
6. obtuse.
7. per unit length on one side of a line in the surface.
8. free surface energy.
9. C.
10. B.
11. D.
12. A.
13. $1\cdot4\times10^{-2}, 8\cdot68\times10^{-3}\,N$.
14. MT^{-2}, (i) 6·6, (ii) 5·5 cm.
15. 2·7 cm.
16. (i) $1\cdot00056\times10^5\,N\,m^{-2}$, (ii) 0·14 cm.
19. 0·75 cm.
20. 2·6 cm.
21. $8\pi T(b^2-a^2)$.
22. 7·35 cm, angle of contact now 60°.
25. $\pi r^2 gh\rho g - 2\pi r\gamma\cos\alpha$.
26. 5·6 cm, angle of contact 44°.
27. 0·18 cm.
28. (a) 3·2 cm, (b) $7\cdot1\times10^{-2}\,N\,m^{-1}$.
30. 0·8.
31. $7\cdot4\times10^{-3}\,N$.

EXERCISES 6 (p. 196)

1. tensile.
2. $N\,m^{-2}$.
3. elastic limit.
4. ½ load.
5. pressure.
6. shear.
7. C.
8. B.
9. A.
10. D.
11. $6\cdot2\times10^6\,N\,m^{-2}, 6\times10^{-5}, 1\cdot0\times10^{11}\,N\,m^{-2}$.
12. 6·8 kg.
13. 0·08 mm.
15. 7·13 kgf.
16. 1·2 N.
17. $20\,m\,s^{-1}$.
18. 117 gf.
19. 2%.
20. 1/120 J.
21. $1\cdot1\times10^4\,J$.
22. (a) $2\cdot2\times10^{-6}\,m$, (b) $5\cdot46\times10^{-5}\,J$.
23. 0·5.
24. 144°C, 3360 kg.
25. (a) $2\times10^{11}\,N\,m^{-2}$. (b) $4\cdot7\times10^{-2}\,J$.
26. $1\cdot1\times10^4\,N$.
27. $7\cdot8\times10^{10}\,N\,m^{-2}$.
28. 40, 74 N.
29. 28·6 kgf.
30. $12\times10^7\,N\,m^{-2}$, 3 kgf.

EXERCISES 7 (p. 219)

1. F/R.
2. independent.
3. $\pi p a^4/8\eta l$.
4. $ML^{-1}T^{-1}$.
5. $6\pi\eta a v$.
6. volume.
7. C.
8. B.

232 MECHANICS AND PROPERTIES OF MATTER

9. D.
11. 4·97 cm.
13. 0·6.
15. (a) 46, (b) 60 rev min^{-1}.
18. $x = 2, y = 1, z = 2; k = 2\cdot3$.
21. 15 h 36 m.
26. $2\pi\eta a_2^3 l\omega/(a_1-a_2)$.

10. A.
12. (a) 17°, (b) 0·36 N.
14. 160 J.
17. 0·5, 13·5 cm.
20. 0·025 m s^{-1}.
24. 17 s.

REVISION PAPERS (p. 223)

A. **2.** 42·4 cm, 2·7%.
4. 2·16 mm; 115 N, 54 N.
6. 16 cm, 80 g.

3. 0·28%, 742·1 mm.
5. 0·39 cm s^{-1}.
7. $\sqrt{3}:2$.

B. **1.** $2\cdot6 \times 10^{-21}$ kg m s^{-1}, 90° to electron direction.
5. $2\cdot6 \times 10^{-2}$ kg m^2.

2. $1\cdot58 \times 10^{-4}$ J, 0.
4. 20 m s^{-1}.
6. $M/2A\rho$; $M/A\rho$; $3M/8\rho A^2$.

C. **1.** (a) 9N, (b) 7·5 N, (c) 4·5 N; (a) 37 N, (b) zero.
2. (a) 1·4 seconds, (b) 0·25 N.
3. (a) $2\cdot13 \times 10^{11}$ N m^{-2}, (b) 0·0249 J.
4. $g_0/g_h = (R+h)^2/R^2$. $V_1 = 2,400$, $V_2 = 7,800$ m s^{-1}.
5. 5·54 m s^{-1}, 79%.
7. (a) 1·01 m, (b) $4\cdot3 \times 10^{-2}$ kg m^2.

D. **1.** $v = kwr^{-1}\eta^{-1}$.
4. 16 g; $3\cdot925 \times 10^{-3}$ J.
7. $2\cdot42 \times 10^{-2}$ N m^{-1}.

3. 36%.
5. $9\cdot315 \times 10^{-4}$ kg m^2.
6. (c) $(T_1-T_2)/200$.

INDEX

Acceleration, 5
  angular, 40
  in circle, 41
  of gravity, 7, 46, 75
  of liquid, 136
  uniform, 6
  units of, 5
Adhesion, 154
Adiabatic modulus, 190
Amplitude, 51
Angle of contact, 153–4
Angular momentum, 94
  conservation of, 94, 97
  velocity, 40
ARCHIMEDES' principle, 132
Atmospheric pressure, 127–30

Balance, common, 122
Banking of track, 45
Bar, 128
Barometer, FORTIN, 128
  corrections to, 128
BERNOULLI'S principle, 138
BOYS, C.V., 68
Breaking stress, 182
Brittle, 182
Brownian motion, 131
Bulk modulus, 189
  of gases, 190
Buoyancy correction, 123

Capillarity, 152, 154, 161
CAVENDISH, 68
Centre of gravity, 117
  of mass, 26, 118
Centripetal force, 42
Centrifuge, 43
Circle, motion in, 40–7
Coefficient of kinetic friction, 201
  of restitution, 23
  of static friction, 200
  of viscosity, 204
Cohesion, 154
Common balance, 122
Components of force, 9, 27
  of g, 9
  of velocity, 8
  resolved, 9

Compound pendulum, 103
Conical pendulum, 48
Conservation of energy, 31
  of momentum, 20
Conservative forces, 30
Constant of gravitation, 68
Couples, 92, 115
  work done by, 116
Critical velocity, 207
Cylindrical surface, 168

Damped oscillations, 53
Density, 132
  determinations of, 126, 134
  of earth, 70
Diffraction, 66
Dimensions, 33
  applications of, 33, 191, 205, 212
Dislocation, 182
Distance-time curve, 3
Drops, 149
Ductile, 182
Dyne, 15

Earth, density of, 70
  mass of, 70
EINSTEIN'S law, 32
Elastic collisions, 22
Elastic limit, 181
Elasticity, 180
Energy in wire, 186
  kinetic, 28
  potential, 29
  rotational, 86, 98
Equilibrium, conditions of, 114
Erg, 27
Errors of balance, 122
Excess pressure in bubble, 158
Explosive forces, 24, 28

Falling drop, 164
  sphere, viscosity by, 213
Fluid motion, 137
Floating body, 135
Flow, orderly, 204
  turbulent, 207
Force, 14–18
  in heated bar, 185
  units of, 14

INDEX

Friction, coefficient of static, 200
  kinetic (dynamic), 201
  laws of, 202
  limiting, 200

Gaseous state, 147
Gravitation, NEWTON's law of, 68
Gravitational constant, 68
  mass, 71
  potential, 78
Gravity, acceleration of, 7, 46, 75
  motion under, 7
Gyration, radius of, 90

HOOKE's law, 181
Horse-power, 27
Hydrometer, 135

Impulse, 17
Inelastic collision, 22
Inertia, 13
  moments of, 86
Inertial mass, 71
Interference, 65

JAEGER's method, 167
Joule, the, 6

KEPLER's laws, 66
Kilogramme force, 15
Kinetic energy, 28

Laminar flow, 204
Latent heat, 170
Limit, elastic, 181
Liquid state, 146
Longitudinal wave, 63

Mass, 13
  energy relation, 32
Mean free path, 217
Modulus of elasticity, adiabatic, 190
  bulk, 189
  isothermal, 190
  of gases, 190
  of rigidity, 191
  YOUNG's modulus, 182
Molecular forces, 145
  sphere of attraction, 149
Moment of couple, 115
  of force, 112
  of inertia, 86
  of cylinder, 88
  of disc, 88
  of flywheel, 100

  of plate, 102
  of ring, 87
  of rod, 87
  of sphere, 89
  theorems of, 90, 91
Momentum, angular, 94
  conservation of angular, 94
  conservation of linear, 19–21
  linear, 17
Moon, motion of, 67
Motion, NEWTON's laws of, 13
  in circle, 40
  in straight line, 1
  of projectile, 9
  simple harmonic, 49
  under gravity, 7

Neutral equilibrium, 121
NEWTON, 13, 66, 204
Newton, the, 14
Non-conservative forces, 31

Oscillations, energy exchange, 69
  damped, 53
  free, 53
Oscillation of liquid, 59
  of rigid body, 101
OSTWALD viscometer, 211

Parallel axes, theorem of, 90
  forces, 113
Parallelogram of forces, 110
  of velocities, 7
Parking orbit, 72
Pascal, the, 126
Pendulum, compound, 103
  simple, 55
Period, 50
Pipe, flow through, 205
Planetary motion, 66
PLATEAU's spherule, 150
Poise, 205
POISEUILLE formula, 205, 208
POISSON's ratio, 194
Potential energy, 29, 78
Power, 27
Pressure, atmospheric, 127
  of liquid, 124
Progressive wave, 63
Projectiles, 9

Radius of gyration, 90
Range, horizontal, 10
Relative density, 132
  velocity, 11

INDEX

Resolution of forces, 111
Retardation, 6
REYNOLDS, O., 207
Rigid body motion, 86
Rigidity, modulus of, 191
Rocket motion, 24
Rod, moment of inertia of, 87
Rolling object, 98
Rotation about axis, 86

Scalars, 12
Sensitivity of balance, 122
Shear strain, 192
    stress, 192
Simple harmonic motion, 49–61
    pendulum, 55
    due to gravity, 77
    energy exchanges, 59
SI units, 14
Soap film, 157
Solid state, 145
Specific gravity, 132
    methods for, 134
Speed, 2
Spring, spiral, 57
Stable equilibrium, 120
Stationary wave, 64
STOKES' law, 212
Strain, bulk, 189
    shear, 191
    tensile, 182
    torsional, 192
Stratosphere, 130
Streamlines, 137
Stress, bulk, 189
    shear, 191
    tensile, 138
    torsional, 192
Stretched wire, energy in, 186
Surface energy, 170–3
Surface tension, 145
    and excess pressure, 158
    dimensions of, 151
    measurement of, 154–7, 160
    theory of, 149
    units of, 150
    variation of, 167

Terminal velocity, 212
Theorem of parallel axes, 90
    of perpendicular axes, 91

Torque, 92, 115
Torricellian vacuum, 104
Torricelli's theorem, 140
Torsion, 193
Torsional vibration, 102
Transverse wave, 64
Triangle of forces, 111
Turbulent flow, 207

Uniform acceleration, 6
    velocity, 1
Unstable equilibrium, 120
Upthrust, 133

VAN DER WAALS, 147
Variation of g, 46, 75
Vectors, 1
Velocities, addition of, 10
    subtraction of, 11
Velocity, angular, 40
    critical, 207
    of escape, 79
    relative, 11
    terminal, 212
    — time curve, 3
    uniform, 1
Viscosity, 203
    determination of, 163–6, 194
    dimensions of, 204
    falling sphere and, 212
    of gas, 216
    molecular theory, 216
    of gas, 216
    variation of, 212

Watt, 27
Waves, 62
    velocity of, 64
Weighting correction, 122–3
Weight, 15
Weightlessness, 74, 137
Wetting of surface, 153
Work, 26
    done by couple, 116
    in stretching wire, 187

Yield point, 182
YOUNG's modulus, 182
    determination of, 184
    dimensions of, 183
    units of, 183